Power Electronics for Electric Vehicles and Energy Storage

This text will help readers to gain knowledge about designing power electronic converters and their control for electric vehicles. It discusses the ways in which power from electric vehicle batteries is transferred to an electric motor, the technology used for charging electric vehicle batteries, and energy storage. The text covers case studies and real-life examples related to electric vehicles.

The book

- Discusses the latest advances and developments in the field of electric vehicles
- Examines the challenges associated with the integration of renewable energy sources with electric vehicles
- Highlights basic understanding of the charging infrastructure for electric vehicles
- Covers concepts including the reliability of power converters in electric vehicles, and battery management systems.

This book discusses the challenges, emerging technologies, and recent development of power electronics for electric vehicles. It will serve as an ideal reference text for graduate students and academic researchers in the fields of electrical engineering, electronics and communication engineering, environmental engineering, automotive engineering, and computer science.

Power Electronics for Electric Vehicles and Energy Storage
Emerging Technologies and Developments

Edited by Dharavath Kishan,
Ramani Kannan, B Dastagiri Reddy,
and Prajof Prabhakaran

CRC Press is an imprint of the
Taylor & Francis Group, an **informa** business

Cover image: © Shutterstock

First edition published 2023
by CRC Press

6000 Broken Sound Parkway NW, Suite 300, Boca Raton, FL 33487-2742

and by CRC Press
4 Park Square, Milton Park, Abingdon, Oxon, OX14 4RN

CRC Press is an imprint of Taylor & Francis Group, LLC

© 2023 selection and editorial matter, Dharavath Kishan, Ramani Kannan, B Dastagiri Reddy and Prajof Prabhakaran; individual chapters, the contributors

Reasonable efforts have been made to publish reliable data and information, but the author and publisher cannot assume responsibility for the validity of all materials or the consequences of their use. The authors and publishers have attempted to trace the copyright holders of all material reproduced in this publication and apologize to copyright holders if permission to publish in this form has not been obtained. If any copyright material has not been acknowledged please write and let us know so we may rectify in any future reprint.

Except as permitted under U.S. Copyright Law, no part of this book may be reprinted, reproduced, transmitted, or utilized in any form by any electronic, mechanical, or other means, now known or hereafter invented, including photocopying, microfilming, and recording, or in any information storage or retrieval system, without written permission from the publishers.

For permission to photocopy or use material electronically from this work, access www.copyright.com or contact the Copyright Clearance Center, Inc. (CCC), 222 Rosewood Drive, Danvers, MA 01923, 978-750-8400. For works that are not available on CCC please contact mpkbookspermissions@tandf.co.uk

Trademark notice: Product or corporate names may be trademarks or registered trademarks and are used only for identification and explanation without intent to infringe.

Library of Congress Cataloging–in–Publication Data
Names: Kishan, Dharavath, editor. | Kannan, Ramani, editor. | Reddy, B. Dastagiri, editor. | P., Prajof, editor.
Title: Power electronics for electric vehicles and energy storage : emerging technologies and developments / edited by Dharavath Kishan, Ramani Kannan, B. Dastagiri Reddy, and Prajof P.
Description: First edition. | Boca Raton, FL : CRC Press, [2023] | Includes bibliographical references and index.
Identifiers: LCCN 2022052983 (print) | LCCN 2022052984 (ebook) | ISBN 9781032164199 (hbk) | ISBN 9781032164243 (pbk) | ISBN 9781003248484 (ebk)
Subjects: LCSH: Electric vehicles--Electronic equipment. | Electric current converters. | Battery chargers. | Power electronics.
Classification: LCC TL220 .P68 2023 (print) | LCC TL220 (ebook) | DDC 629.22/93--dc23/eng/20230110
LC record available at https://lccn.loc.gov/2022052983
LC ebook record available at https://lccn.loc.gov/2022052984

ISBN: 978-1-032-16419-9 (hbk)
ISBN: 978-1-032-16424-3 (pbk)
ISBN: 978-1-003-24848-4 (ebk)

DOI: 10.1201/9781003248484

Typeset in Sabon
by Deanta Global Publishing Services, Chennai, India

Contents

Preface	*vii*
Editor biographies	*ix*
List of contributors	*xiii*

1 Introduction to hybrid electric vehicle systems   1
VINUKUMAR LUCKOSE, RAMANI KANNAN, AND DHARAVATH KISHAN

2 Onboard battery charging infrastructure for electrified transportation   27
HARISH KARNEDDI AND DEEPAK RONANKI

3 Design and development of LLC resonant converter for battery charging   61
AKASH GUPTA AND MAHMADASRAF A MULLA

4 Performance evaluation of multi-input converter-based battery charging system for electric vehicle applications   93
R KALPANA, KIRAN R, AND P PARTHIBAN

5 Analysis of solar pv-based electric vehicle charging infrastructure with vehicle-to-grid power regulation   117
SHEIK MOHAMMED SULTHAN, MOHAMMED MANSOOR O, AND ULAGANATHAN M

6 Wireless power transfer-based next-generation electric vehicle charging technology   137
PRIYANKA TIWARI AND DEEPAK RONANKI

vi *Contents*

7 Asymmetric clamped mode control for output voltage
   regulation in wireless battery charging system for EV   175
   DHARAVATH KISHAN, MARUPURU VINOD, B DASTAGIRI REDDY,
   AND RAMANI KANNAN

8 Selection of electric drive for EVs with emphasis on switched
   reluctance motor   193
   PITTAM KRISHNA REDDY, P PARTHIBAN, AND R KALPANA

9 Voltage lift quasi Z-source inverter topologies for
   electric vehicles   217
   JOSEPHINE R L, GOLLAPINNI VAISHNAVI, AKSHATHA PATIL,
   M SAI HARSHA NAIDU, AND RACHAPUTI BHANU PRAKASH

10 Sensorless rotor position estimation and regenerative braking
   capability of a solar-powered electric vehicle driven by PMSM 245
   VIJAYAPRIYA R, ARUN S L, AND RAJA PITCHAMUTHU

11 Performance analysis of the integrated dual input converter
   for EV battery charging application   271
   KUDITI KAMALAPATHI AND PONUGOTHU SRINIVASA RAO NAYAK

   Index   301

Preface

Energy is a crucial resource that all countries around the globe are pursuing to ensure their citizens' social and economic wellbeing. Most nations around the world, however, face innumerable challenges in procuring adequate, cheap, and suitable supplies of energy sources. For countries that are heavily dependent on energy imports, there are always threats of energy disruptions and high prices of energy resources due to political and economic issues occurring in any part of the world. Furthermore, there are also concerns related to environmental pollution and climate change arising from the burning of fossil fuels. In a world where energy/fuel security and environmental protection are growing concerns, the development of electric and hybrid electric vehicles is increasing rapidly.

Electric vehicles (EVs) have relatively high well-to-wheel efficiency, hence providing higher fuel economy and significantly reducing environmental pollution. Despite being a century-old concept, EVs were not popular in the past due to high cost, limited performance, short range, and long charging times. However, with advancements in power electronics, electric motors, and the invention of high-performance batteries, the performance of an EV has drastically increased and outpowered a conventional internal combustion engine-based vehicle in terms of performance. In an EV, power electronics mainly process and control the flow of electrical energy from the energy source (i.e., battery, fuel cell, hybrid storage, etc.) to the motor. Power electronics form the core component in plug-in charging. They also enable fast and wireless charging of EV batteries. Today, EV's power electronics utilize wide band gap (WBG) semiconductor devices (like silicon carbide [SiC] and gallium nitride [GaN]) for efficient control and conversion of electric power. These WBG devices can withstand higher temperatures and operate at high frequencies with low power losses. Because of these features, the WBG-based power converter offers high efficiency and power density. Today, research on power converters for EV applications focuses on novel topologies and designs that can reduce part counts and enable reliable and modular components. Research on fast charging, renewable energy integrated solutions, charging infrastructure, high-performance and viii *Preface*

low-cost motor drive algorithms, wireless charging, etc., is gaining momentum for EV applications.

In this book, we have attempted to include recent research findings, challenges, emerging technologies, and recent developments in the application of power electronics to electric vehicles. The major topics covered in this book include: (i) introduction to electric and hybrid electric vehicles (Chapter 1), (ii) onboard charging infrastructure (Chapter 2), (iii) power converter topologies for EV applications (Chapters 3 and 4), (iv) integration of renewable energy for EV and energy storage applications (Chapter 5), (v) recent advancements in wireless charging (Chapters 6 and 7), and (vi) electric drives, algorithms, and traction control for EVs (Chapters 8, 9, 10, and 11).

We want to acknowledge the efforts of various authors who have contributed to this book and the assistance of the staff of the CRC Press, Taylor & Francis team to make this book a reality.

Editor biographies

Dharavath Kishan received the BTech degree in Electrical and Electronics Engineering and MTech degree in Power Electronics from Jawaharlal Nehru Technological University Hyderabad in 2011 and 2013, respectively, and he received his PhD degree from National Institute of Technology Tiruchirappalli in 2018. Currently, he is working as Assistant Professor in the Department of E & E Engineering at the National Institute of Technology Karnataka, Surathkal, India. Prior to joining the National Institute of Technology Karnataka, he worked as Assistant Professor in the Faculty of Science and Technology, ICFAI Foundation for Higher Education, Hyderabad. Dr Kishan's current research interests include power electronics and its applications in electric vehicles, wireless power transfer and transportation electrification. He has published 22 research papers in reputed journals and peer reviewed international conferences. He has also delivered guest lectures at various events on wireless power transfer for electric vehicles. He is also an Institute of Electrical and Electronics Engineers senior member and an Industry Applications Society member. He has guided five master's level students in the area of power electronics and is currently guiding two master's and six PhD students in the area of power electronic applications in electric vehicles.

Ramani Kannan received the BEng degree from Bharathiyar University, Coimbatore, India, and the MEng and PhD degrees in power electronics and drives from Anna University, Chennai, India. He is currently a Senior Lecturer with Universiti Teknologi PETRONAS, Seri Iskandar, Malaysia. He holds more than 135 publications in reputed international and national journals and conferences. His research interests include power electronics, inverters, modelling of induction motors, artificial intelligence, machine learning, and optimization techniques. Dr. Kannan is a Chartered Engineer (CEng), UK, a member of the Institution of Engineers (India), and a senior member of the Institute of Electrical and Electronics Engineers, the Institution of Engineering and Technology (UK), the Indian Society for Technical Education, and the x *Editor biographies*

Institute of Advanced Engineering and Science. He has been the recipient of numerous awards related to research and teaching, and grants from the Fundamental Research Grant Scheme, the Association of Southeast Asian Nations-India, Yayasan Universiti Teknologi PETRONAS, the Ministry of Energy, Green Technology & Water (Malaysia), and the Sexually Transmitted Infection Research Foundation. He is an Associate Editor for *IEEE Access* and *International Transactions on Electrical Energy Systems* (Wiley). He was the Editor-in-Chief for the *Journal of Asian Scientific Research* from 2011 to 2018. He is also a Regional Editor for the *International Journal of Computer Aided Engineering and Technology*, UK, and an *Associate Editor for Advanced Materials Science and Technology*.

B Dastagiri Reddy received the BTech degree in electrical and electronics engineering and the MTech degree in power electronics from Jawaharlal Nehru Technological University, Anantapur, India, in 2009 and 2011, respectively, and the PhD degree in power electronics and its applications to renewable energy from National Institute of Technology, Tiruchirappalli, India, in 2015. From 2015 to 2018, he worked as Research Fellow with the Department of Electrical and Computer Engineering, National University of Singapore, Singapore. He also worked as Scientist with ABB Global Industries and Services Limited, Chennai, India. Currently, he is working as Assistant Professor in the Department of E & E Engineering at National Institute of Technology Karnataka, Surathkal, India. His research interests include power electronics, renewable energy, micro-grids, electric vehicles, marine propulsion, high-voltage direct current, and flexible alternating current transmission systems.

Prajof Prabhakaran received the BTech degree in Electrical & Electronics Engineering from Amrita Vishwa Vidyapeetham (Amrita University), Coimbatore, in 2009, followed by the MTech degree in Power Electronics from the same university in 2011. He received the PhD degree in Electrical Engineering from the Indian Institute of Technology Bombay, Mumbai, in 2018. His doctoral research was focused on direct current micro-grids, and he has published four Institute of Electrical and Electronics Engineers transactions and five Institute of Electrical and Electronics Engineers international conferences from his research work. From July 2011 to June 2013, he worked as an Assistant Professor in the Electrical & Electronics Engineering department, Amrita School of Engineering, Coimbatore. From July 2013 to June 2018, he worked as a teaching assistant and a lab coordinator at the Applied Power Electronics Lab, Indian Institute of Technology Bombay. After a brief stint with National Institute of Technology Calicut, Kozhikode, as an Ad-hoc Faculty during July–November 2018, he joined the Transportation Solutions department, L&T Technological Services, Bangalore, where he worked for a year as a Project Lead for the research and development of powertrains and battery management systems for electric vehicles. On 4 December 2019, he joined the National Institute of Technology Karnataka, Surathkal, where he is currently an Assistant Professor in the Department of Electrical and Electronics Engineering. He mainly works in the field of power electronics with a focus on renewable energy source-based systems, microgrids, and electric vehicles. Dr. Prajof is a reviewer for *IEEE Access, IEEE Transactions on Power Electronics, IEEE Transactions on Industrial Electronics, IEEE Transactions on Smart Grid, and IEEE Transactions on Industry Applications.*

List of contributors

Sushanta Bordoloi
National Institute of Technology Mizoram
Mizoram, India

Saransh Chourey
Energy Institute Bengaluru
Centre of Rajiv Gandhi Institute of Petroleum Technology
Bengaluru, India

Akash Gupta
Sardar Vallabhbhai National Institute of Technology
Surat, India

R Kalpana
National Institute of Technology Karnataka
Surathkal, India

Kuditi Kamalapathi
National Institute of Technology Tiruchirappalli
Tiruchirappalli, India

Ramani Kannan
Electrical and Electronics Engineering
Universiti Teknologi PETRONAS
Seri Iskandar
Perak, Malaysia

Harish Karneddi
Indian Institute of Technology Roorkee
Roorkee
Uttarakhand, India

Dharavath Kishan
National Institute of Technology Karnataka
Surathkal, India

xiv *List of contributors*

Josephine R L
National Institute of Technology Tiruchirappalli
Tiruchirappalli, India

Arun S L
Vellore Institute of Technology
Vellore
Tamil Nadu, India

Vinukumar Luckose
SEGi University
Petaling Jaya, Malaysia

Ulaganathan M
PSN College of Engineering
Tirunelveli
Tamil Nadu, India

Mahmadasraf A Mulla
Sardar Vallabhbhai National Institute of Technology
Surat, India

M. Sai Harsha Naidu
National Institute of Technology Tiruchirappalli
Tiruchirappalli, India

Ponugothu Srinivasa Rao Nayak
National Institute of Technology Tiruchirappalli
Tiruchirappalli, India

Mohammed Mansoor O
TKM College of Engineering
Kollam
Kerala, India

P Parthiban
National Institute of Technology Karnataka
Surathkal, India

Akshatha Patil
National Institute of Technology Tiruchirappalli
Tiruchirappalli, India

Raja Pitchamuthu
National Institute of Technology Tiruchirappalli
Tiruchirappalli, India

Rachaputi Bhanu Prakash
National Institute of Technology Tiruchirappalli
Tiruchirappalli, India

List of contributors xv

Kiran R
National Institute of Technology Karnataka
Surathkal, India

Vijayapriya R
Vellore Institute of Technology
Vellore
Tamil Nadu, India

B Dastagiri Reddy
National Institute of Technology Karnataka
Surathkal, India

Pittam Krishna Reddy
National Institute of Technology Karnataka
Surathkal, India

Deepak Ronanki
Indian Institute of Technology Madras
Chennai
Tamil Nadu, India

Sheik Mohammed S
Universiti Teknologi Brunei
Bandar Seri Begawan, Brunei Darussalam

Priyanka Tiwari
Indian Institute of Technology Roorkee
Roorkee
Uttarakhand, India

Gollapinni Vaishnavi
National Institute of Technology Tiruchirappalli
Tiruchirappalli, India

Marupuru Vinod
National Institute of Technology Karnataka
Surathkal, India

1 Introduction to hybrid electric vehicle systems

Vinukumar Luckose, Ramani Kannan, and Dharavath Kishan

1.1 Introduction

The automobile industry has made a tremendous change in modern society by fulfilling people's mobility needs. One of the greatest achievements in the automobile industry was the development of internal combustion engine (ICE) vehicles. This has led to greater use of vehicles in human society. However, the use of ICE vehicles creates severe environmental pollution worldwide. One effective way to reduce environmental pollution is to increase the efficiency of energy conversion by developing renewable and alternative energy sources [1]. The invention of electric vehicles (EVs) is helping the automobile industry to overcome these environmental issues.

This chapter explains the historical development of EV systems and provides an overview of EV drive technologies and hybrid electric vehicles (HEVs), various hybrid electric drive technologies with their operation modes, and power train concepts with the level/degree of hybridization and the types of HEV based on the nature of power sources.

1.2 What is an electric vehicle?

An EV is a transportation system that uses electric motors or the combination of electric motors and the ICE. This system leads towards the reduction of hazardous emissions of fuel gas, improvement of vehicle performance, increased vehicle efficiency, and reduced consumption of fossil fuel [2]. EV systems comprise four main drive technologies. These are fuel cell electric vehicles (FCEVs), battery electric vehicles (BEVs), HEVs, and plug-in HEVs (PHEVs).

1.3 Historical development of electric vehicles

In 1769, Nicolas-Joseph Cugnot designed the first steam-powered car, which was constructed by M. Brezin with heavy materials and was able to travel on a perfectly flat surface. The ICE car was developed by Isaac de Rivaz in 1807. The first electrical car was developed by Robert Anderson of Aberdeen

DOI: 10.1201/9781003248484-1

2 Power electronics for electric vehicles

in 1839. In 1900, steam-powered cars were manufactured more in the industry due to their high efficiency in terms of power and speed. However, their poor fuel economy, the need to heat the boiler before driving and the loss of fresh water for steam operation led the automobile market to the development of gasoline cars [3]. Even though gasoline cars provided better power and utility, they were very noisy and unreliable [4] in operation. This led to the invention of electric cars. Electric cars are noiseless, relatively easy to operate, comfortable, clean, and easy to control, produce no toxic emissions, and are fashionable. At the beginning of 1900, electric cars were run using lead acid batteries. From 1890 to 1924, the production of electric cars [1] increased, marking the golden age of EVs, with the highest production in 1912. However, due to lower energy storage and the need for frequent battery recharging, EV production was dropped at the end of 1924. After the downfall of EV production, there was a high demand for gasoline cars in the automobile market from 1924 to 1960. The number of gas stations increased with the shortage of gasoline in the 1970s. During this situation, engineers realized that a gasoline engine combined with an electric motor could be good for future automobiles to produce modern hybrid cars [1].

The historical development of EVs in four different periods is shown in Table 1.1. The early development of electric cars was between 1828 to 1867, called the pre–electric car age. The golden age of the EV industry was from 1881 to 1910. The production of electric cars decreased tremendously from 1910 to 1969, the dark age in EV history. The regenerative braking concept was developed in 1978, and this has caused the vehicle industry to move towards the modern electric car age.

1.4 Electric vehicle drive technologies, configurations, and comparison

Based on the degree of electrification as shown in Figure 1.1, there are four different electric drive technologies to control EVs. These are the FCEV, BEV, HEV, and plug-in hybrid electric vehicle (PHEV) technologies [7].

1.4.1 Fuel cell electric vehicles (FCEVs)

A fuel cell is a device that generates electrical energy by combining hydrogen and oxygen. When hydrogen and oxygen are combined together by chemical reactions, the products are water, electricity, and heat. The type of fuel cell used for vehicle and fuel cell applications is the proton-exchange membrane (PEM) [8]. The reactants in fuel cells are not harmful to the environment, as only water is produced during the chemical reaction [9]. FCEVs are classified into conventional FCEVs and hybrid fuel cell vehicles. Figure 1.2 shows a conventional FCEV system in which the refined hydrogen gas is stored in the fuel tank. The hydrogen gas will pass through the fuel cell stack, come directly into contact with the oxygen in the atmosphere, and go through a chemical reaction that produces electrical energy [10]. The electrical energy

Hybrid electric vehicle systems   3

Table 1.1 Historical Development of Electric Vehicles

Period	Year	Description
Pre–Electric Car Age	1828	Small toy electric vehicle built by Anyos Jedlik using a motor
	1834	Battery electric vehicle developed by Thomas Devenport
	1839	First electric vehicle built by Robert Anderson of Aberdeen
	1867	Electrically powered bicycle invented by Franz Kravogl [4]
Golden Age	1881	Electrically powered tricycle invented by Gustave Trouve
	1884	Thomas Parker developed an electric car with a high-capacity rechargeable battery
	1888	Four-passenger carriage with 1-horse power motor and 24-cell battery built by Immisch & Company
	1888	Electric car with three wheels developed by Magnus Volk
	1890	First hybrid car, the Lohner-Porsche Elektromobil, created by Ferdinand Porsche [1]
	1895	Six-passenger electric car built by William Morrison
	1896	Production of electrical starter motor for gasoline engines
	1898	The first-ever speed record set in an electric car [4]
	1900	Porsche developed the series hybrid car [1]
	1910–1920	End of electric vehicle manufacturing due to large production of petroleum and crude oil
Dark Ages	1920–1950	Limited use of electrical vehicles in the automobile industry [4]
	1915	Parallel hybrid vehicle production by Woods motor vehicle manufacturers
	1959	Development of a "self-charging" battery-powered car
	1960	Numerous hybrid cars designed by Victor Wouk, known as "Godfather of the Hybrid"
Modern Electric Car Age	1978	David Arthurs, an electrical engineer, developed the concept of a regenerative braking system
	1990	Modern hybrid cars developed
	1993	Numerous government agencies and many car manufacturers developed HEVs using alternate energy sources
	2005	Ford car company released Hybrid SUV in 2005
	2008	Tesla Roadster, a battery-operated electric vehicle sports car that travelled more than 200 miles per charge, developed [5]
	2009	Charging station infrastructure began globally
	2009	The Elantra LPI was one of the first electric hybrid cars, run by an ICE fuelled by LPG and lithium-ion polymer batteries [6]
	2018	Tesla Roadster electric sports car launched into space as dummy payload, carried by a Falcon Heavy test flight

4 Power electronics for electric vehicles

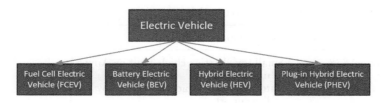

Figure 1.1 Classification of different types of EV.

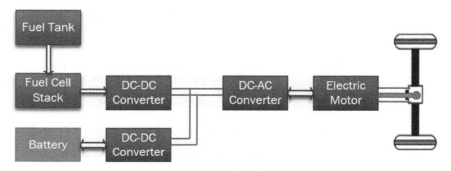

Figure 1.2 Configuration of FCV.

is then supplied to a dc–dc converter, which converts the fixed direct current (dc) to variable dc. Furthermore, the electrical power is supplied to a dc–(ac) converter that converts the direct current to alternating current (ac) to drive the electrical motor. In addition to the fuel cell, a rechargeable battery is used to supply the extra power to the vehicle during short acceleration and when the fuel cell is off. The rechargeable battery can be charged during braking or can be charged externally. Fuel cell technology is attracting special attention among the other various renewable energy sources because of its ability to deliver an incredible amount of energy, its high efficiency, and its zero tailpipe emissions. However, fuel cell technology has some limitations: high cost, lack of fuel cell infrastructure, and it is expensive to install the fuelling stations [2].

1.4.2 Battery electric vehicles (BEVs)

BEVs are popular nowadays because of their impact on reducing environmental pollution. These vehicles are popularly called EVs that use rechargeable batteries instead of using an ICE [11]. The configuration of a BEV is shown in Figure 1.3. It consists of a rechargeable battery, a dc–ac converter, an ultra-capacitor, a dc–dc converter, and an electric motor. The

Hybrid electric vehicle systems 5

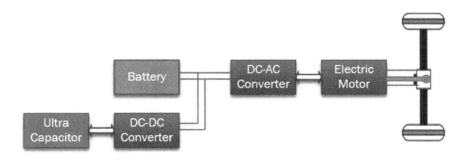

Figure 1.3 Configuration of BEV.

rechargeable battery needs to be charged externally from a charging station, and it supplies the charge to the dc–ac converter. This converter helps to control the speed and torque regulation of the electric motor. The electric motor produces the necessary mechanical power to drive the wheel through the transmission system. The ultra-capacitor in the BEV reduces the loss in the battery in order to increase the efficiency. This capacitor stores energy during regenerative braking and delivers energy during peak acceleration.

Advantages of BEVs are:

- Less pollution because there is no exhaust system
- Cheaper to run and maintain
- Quieter and more comfortable to drive.

Disadvantages of BEVs are:

- Short ranges for driving
- Longer time to charge the battery
- High car price
- Lack of charging stations and less choice [12].

1.4.3 Plug-in hybrid electric vehicles (PHEVs)

PHEVs have two power sources: one is a large rechargeable battery with an electric motor, and the other uses fuel such as gasoline to power an ICE. The rechargeable batteries can be charged with the help of a charging station or by the concept of a regenerative mode of braking [13]. The battery will supply the entire driving power to the transmission system until the battery is exhausted, and it will then automatically run with the aid of ICE. Figure 1.4 shows the configuration of a PHEV in which the ICE and the electrical system with battery, converter, and electric motor are coupled to the drive system. The power flow from the battery to the electric motor is

6 Power electronics for electric vehicles

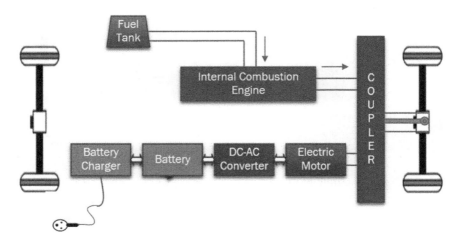

Figure 1.4 Configuration of PHEV.

controlled by the bidirectional converter. These converters also manage the power flow when the ICE is on or off and provide the necessary torque to control the speed of the motor.

1.4.4 Hybrid electric vehicle (HEV)

HEV technology combines the energy from the electric system and the ICE to drive the vehicle. Figure 1.5 shows the configuration of an HEV system. The power sources that supply energy to the HEV system are the ICE, the battery, and an ultra-capacitor. The battery will be charged with the help of the ICE and by the concept of regenerative braking [2]. The ultra-capacitor

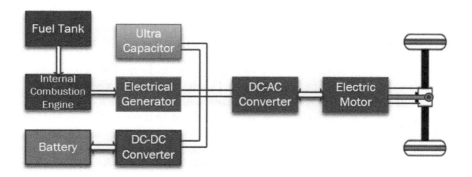

Figure 1.5 Configuration of HEV.

Hybrid electric vehicle systems 7

stores energy during normal driving and releases the energy during peak acceleration. The fuel consumption in HEV technology can be reduced by recapturing energy during braking, downscaling the ICE, and shutting off the ICE when the vehicle is not moving [14].

1.4.5 Comparison of electric vehicle drive technologies

Table 1.2 summarizes the comparison of EV drive technologies based on their drive system, energy storage system, main issues, advantages, and disadvantages. The hybrid electric drive technology has more features than the BEV and FCEV drive technology [15]. The limitations of FCEV are high cost, lack of hydrogen fuel cell infrastructure, and it is expensive to install the fuelling stations. Although the BEV has several advantages, the technology still suffers from limitations to the speed range, charging technology, and charging stations.

1.5 Overview of hybrid electric vehicles (HEVs)

HEVs combine the drive powers of an electrical drive storage system and an ICE to deliver the energy to the vehicle wheel through the traction drive system. Batteries are used to store energy along with the help of an ultra-capacitor to drive the electric motor in the electrical drive system. The ICE and the fuel system are used to generate mechanical power in the fuel-based energy storage system. There are many different types of hybrid electric drive system currently available, employing ICE, solar cell, gas-operated turbines, alternative types of fuel engines, hydrogen-based fuel cells and in conjunction with rechargeable batteries [2].

When the power requirement is low, the HEV utilizes the electric propulsion drive system. Also, the HEV reduces fuel consumption [16] when the ICE stays off during idle operation or when the vehicle stops at traffic signals. Furthermore, the HEV switches to the ICE fuel system when higher speed is needed. Sometimes, the ICE and electric drive system combine together to increase the vehicle's driving performance. The HEV system is able to decrease or eliminate the turbo lag in turbo-geared vehicles by filling the gaps during gear shift and increase the speed when needed. The rechargeable batteries are charged through ICE and also charged during regenerative braking [17]. These enhanced features have led car manufacturers to widely use the HEV configuration. Nowadays, many types of vehicles have adopted the hybrid vehicle technology to reduce the fuel consumption and environmental pollution. These vehicles include hybrid electric buses, hybrid electric trucks, hybrid locomotives, electric boats, helicopters, tractors, and military vehicles.

Table 1.3 summarizes the detailed information on the various main components of HEVs with their functions. In an HEV system, electric traction motors act as a prime component to drive the wheels of the vehicle

Table 1.2 Comparison of Various Electric Drive Technologies

	FCEV	BEV	HEV
Drive system	Electric motor	Electric motor	• Electric motor • ICE
Energy sources	• Hydrogen tank • Rechargeable battery/ultra-capacitor	• Rechargeable battery • Ultra-capacitor	• Battery • Ultra-capacitor • Gasoline or alternative fuels
Advantages	• Lower gas emissions • Energy efficiency is high • Recapturing energy during braking • No gasoline oil required	• Only electrical energy used • Lower gas emissions • Recapturing energy during braking • Operating cost is lower • Noiseless operation	• Fuel cost is lower • Lower fuel consumption • Lower gas emissions • Recapturing energy during braking
Disadvantages	• Very expensive • Lower availability of hydrogen generation • Fewer hydrogen fuelling stations • Difficult for mass manufacturing	• Low mileage • Improvement needed in battery charging technology • Longer time to recharge • Fewer charging stations	• Initial cost is very high • Complexity of two powertrains • Component availability
Main issues	• Fuel cell cost • Hydrogen production • Hydrogen fuel stations	• Battery management • Charging facilities • Cost • Battery life time	• Battery management • Control and management of various energy sources

Table 1.3 Main Components of an HEV

Key components	Function
Auxiliary and traction batteries	Auxiliary battery is used to supply electric power to start the vehicle at the beginning, and the traction battery provides supply power to all other vehicle accessories
dc–dc converter	Converts high voltage from the traction battery to a low voltage, which is necessary to operate the vehicle accessories and also used to recharge the auxiliary battery
Electric traction motor	The motor provides the necessary mechanical power to the vehicle wheel by using the electrical power from the traction battery
Electric generator	Generates electrical power from the wheels during braking and supplies back to traction battery for recharging
Fuel tank	Stores the gasoline in the vehicle for the ICE operation
ICE	The gasoline is injected into the engine chamber and then mixed with air or fuel mixture, which produces gases at high temperature and pressure
Power electronics controller	Used to manage the power flow from the battery to the electric drive system and also control the speed and torque of the motor
Exhaust system	Guides the exhaust gases away from the ICE

and provide full torque, low noise, and high efficiency at low speeds. Furthermore, the HEV's other performance advantages include efficient driving control, outstanding "off the line" acceleration, and providing good tolerance during fault and flexibility in operation during voltage variations [18].

The HEV system uses different types of electrical motors for the drive application. These are the permanent magnet synchronous motor (PMSM), brushless dc motor (BLDC), three phase induction motor, and switched reluctance motor (SRM). These motors are able to function as a generator to regenerate energy during braking. Furthermore, the air-conditioning system and other equipment such as power steering and lighting system are controlled by electric motor instead of the ICE. This permits the HEV to gain more efficiency because the accessories and other equipment can operate at a constant speed or can be turned off regardless of how fast the ICE is running. The battery voltage in the HEV is much higher than in the battery used in the ICE to reduce the amount of current and power losses.

1.5.1 Hybrid electric drive configurations and their operation modes

The combined drive powers of an electrical energy storage device and an ICE provide the necessary energy for the propulsion system and all vehicle

10 Power electronics for electric vehicles

accessory systems [2]. The main components in an HEV, as mentioned in Section 1.4 can be classified into three different groups:

- Drivetrains – These integrate the ICE system and the electric traction drive system.
- Energy storage unit – The energy storage system uses a large number of batteries to store energy for vehicle operation with high power capabilities.
- Power control unit – This manages the operation of the ICE, the electric drive system, and the energy storage unit.

Figure 1.6 shows the general structure of the drivetrain and possible energy flow [19] directions in an HEV system [20]. The following are the possible ways of power flow in an HEV vehicle to meet the driving requirements:

- Power can be delivered to load by powertrain 1 or 2 alone
- Power can be delivered to load by both powertrain 1 and 2 at the same time
- During braking, powertrain 2 obtains power from load
- Powertrain 1 can supply the power to powertrain 2
- Powertrain 2 receives power from powertrain 1 and load at the same time

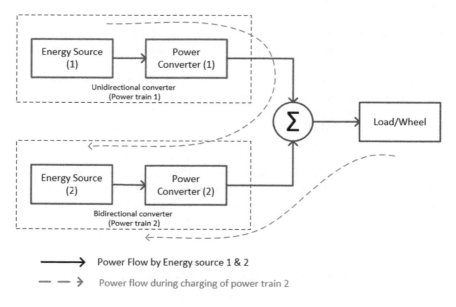

Figure 1.6 General structure of drivetrain.

Hybrid electric vehicle systems 11

- The power can be delivered to powertrain 2 and load by powertrain 1 at the same time
- Powertrain 2 delivers power to load and receives power from powertrain 1
- The load receives power from powertrain 1 and supply to powertrain 2

The various components in an HEV can be combined in different ways to make up the vehicle design configuration. Based on the vehicle design configuration, HEV systems can be classified into four types. These are series hybrid configuration, parallel hybrid configuration, series-parallel hybrid configuration, and complex hybrid configuration

1.5.1.1 Series hybrid configuration
The general structure of the series hybrid configuration of HEV is shown in Figure 1.7. In the series hybrid configuration, the ICE operates the electric generator instead of driving the vehicle wheels directly. The traction drive power can be provided by the ICE and energy stored by the battery. The battery is recharged through the ICE and recaptures the energy during braking.

The series hybrid layout, as shown in Figure 1.8, consists of the ICE and the electric drive systems. The electric generator supplies the necessary voltage to the electric motor to produce the mechanical rotation. The drivetrain system is connected to the electric motor and thus provides the entire driving power to the vehicle wheel. The fuel tank acts as a unidirectional energy source, whereas the ICE operates as a unidirectional energy converter and thus provides the mechanical rotation to the electric generator. During normal operation, the generator supplies the power to the electric motor and the rechargeable batteries. The rechargeable battery is the bidirectional source, and the dc–dc converter acts as the bidirectional energy

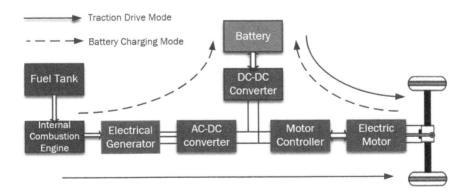

Figure 1.7 General structure of series hybrid vehicle.

12 *Power electronics for electric vehicles*

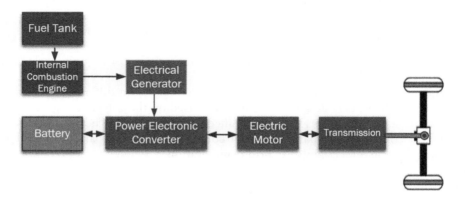

Figure 1.8 Series hybrid vehicle layout.

converter. The power flow between the electric drive system and the battery is controlled by the dc–dc converter. During peak acceleration, the electric motor receives power from the rechargeable batteries and the electric generator [21]. The electric motor can be controlled to operate as a generator and motor and can also run in forward and reverse direction [22]. During braking, the electric motor acts as a generator and supplies the power to charge the battery and ultra-capacitor (if needed). The ultra-capacitor can be used to improve the efficiency of the series hybrid system and minimize the stress factor of the battery.

Advantages of series HEV:

- No mechanical connection between the ICE and the vehicle wheel
- It provides better torque–speed characteristics
- Needs only simple control strategies
- The ICE operates in the most efficient range
- Most efficient during slow driving.

Disadvantages of series HEV:

- Losses may be increased in the generator and motor during the energy conversion
- The cost and weight of the vehicle are higher due to the generator and motor
- The size of the traction motor must meet the maximum requirements.

1.5.1.2 *Parallel hybrid configuration*

Figure 1.9 shows the general structure of a parallel HEV system. This shows the traction drive and battery charging mode of a parallel hybrid

Hybrid electric vehicle systems 13

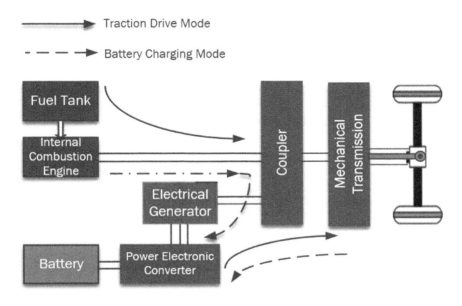

Figure 1.9 General structure of parallel hybrid vehicle.

vehicle. In a parallel hybrid vehicle, both the ICE and the battery supply the traction drive power to the vehicle wheel. The battery can be charged through the ICE and during regenerative braking mode. During braking, the electric motor acts as a generator and supplies electrical energy to the battery [23].

The parallel hybrid layout shown in Figure 1.10 consists of the ICE drive system and the electric drive system. Both systems are coupled to the drive shaft of the wheel through the clutches. Based on the layout shown, the drive power may be supplied by ICE only, or by electric motor, or by both, depending on whether the load requirement is high or low.

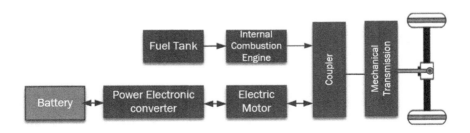

Figure 1.10 Parallel hybrid vehicle layout.

14 *Power electronics for electric vehicles*

Advantages of parallel HEV:

- Energy loss is lower
- Efficiency is higher during long and cruise driving
- Higher flexibility when switching power between electric and ICE
- Electric motor used with lower power compared with series hybrid.

Disadvantages of parallel hybrid electric vehicle:

- The operating points of an engine cannot be fixed properly
- The control strategy and the mechanical configurations are complex compared with the series hybrid drivetrain
- Mostly used for small vehicles due to its compact characteristics.

1.5.1.3 Series-parallel hybrid configuration

Figure 1.11 shows the layout of a series-parallel hybrid vehicle system. This vehicle system can be called the dual or power-split hybrid system. The power is supplied to the wheel by electrical or mechanical means or by both. These vehicles are designed with a suitable control strategy to combine the benefits of series and parallel HEV technologies. By combining the series and parallel hybrid, this system gains an additional mechanical link and an extra generator when compared with the series hybrid and the parallel hybrid type. Even though this configuration has more advantages, it is costly and complex due to the presence of a planetary gear [17] unit, which combines the ICE, the electric generator, and the electric motor.

The system can be made less complex with the use of a trans-motor. In a series-parallel hybrid system, the drive power can be supplied by the ICE only, or the battery only, or both. The ICE supplies the traction power directly to the wheel as well as supplying the mechanical power to the electric

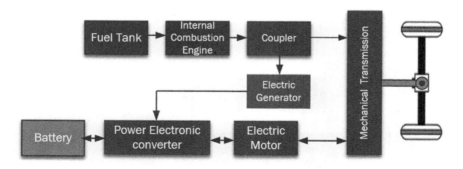

Figure 1.11 Series-parallel hybrid vehicle layout.

generator, thus producing electrical energy. The electrical energy is then supplied to the battery for recharging. The generator in the series hybrid vehicle provides unidirectional power flow from ICE to battery. When the vehicle needs additional drive power, the battery and ICE combine together to supply the drive power to the wheel.

1.5.1.4 Complex hybrid configuration

The complex hybrid drive system shown in Figure 1.12 involves a complex configuration. This configuration is similar to the series-parallel hybrid configuration due to the presence of both an electric generator and a motor [17].

However, the main difference is that the electric generator in a series-parallel hybrid provides unidirectional power flow, whereas the motor in a complex hybrid provides bidirectional power flow. The complex hybrid can also be called a series-parallel configuration based on the current market terminologies. The vehicles have some drawbacks. These are mainly high cost and complexity. But, some vehicles, such as dual axle propulsion systems, have adopted this configuration. In a complex hybrid system, a constantly variable transmission (CVT) system can be used for power splitting between the ICE and the battery to drive the wheels.

Advantages of complex hybrid electric vehicles:

- High flexibility in switching the system between electric and ICE
- Efficient ICE design.

Disadvantages of complex hybrid electric vehicles:

- More complicated and expensive than series and parallel hybrid systems
- Lower efficiency due to multiple electrical conversions.

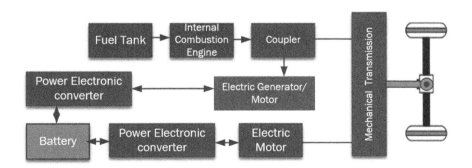

Figure 1.12 Complex hybrid vehicle layout.

16 *Power electronics for electric vehicles*

1.5.2 *Power flow in hybrid electric vehicles*

Because of various electric drive configurations in HEV systems, the vehicle needs a different power control strategy to manage the power flow from or to various components. The main purpose of the control strategy is to increase the fuel efficiency, minimize the emissions, increase the driving performance, and decrease the operating cost. In HEVs, a few parameters need to be considered when designing the power flow control strategies. They are:

- Optimum operating point of the ICE: This control strategy is based on the growth of fuel economy, the reduction of gas emissions, or a negotiation between fuel economy and the gas emissions.
- Optimum operating line of the ICE: In order to deliver the various power demands by the HEV, the corresponding optimal operating points found an optimal operating line.
- Safe battery voltage: During charging, discharging, and regenerative charging, the battery voltage may be altered. During these conditions, the voltage in the battery does not exceed the maximum rated voltage and does not fall below the minimum rated voltage [3].

1.5.2.1 *Power flow mode in series hybrid*

The series hybrid system has four different types of operating mode.

Mode 1 (at normal driving or at acceleration period): Figure 1.13a shows the power flow mode of series hybrid vehicle during normal driving. The ICE and the battery supply the drive power to the vehicle wheel through the electric motor. The power electronics converter controls the power flow between the battery and the electric generator [3, 24].

Mode 2 (at minimum load condition): When the vehicle is running at minimum load, the ICE output will be higher than the power necessary to operate the wheels. During this state, as shown in Figure 1.13b, the required amount of energy is used to drive the traction wheels, and the excess power from the ICE helps to charge the battery through an electric generator and a power converter.

Mode 3 (at braking): Figure 1.13c shows the braking mode of a series hybrid vehicle. When the vehicle undergoes braking or deceleration, the electric motor (acting as a generator) converts the kinetic energy of the wheel into electricity. The energy from the generator is then supplied to the battery for recharging.

Mode 4 (at stopping): Figure 1.13d shows the power flow mode of a series hybrid during vehicle stopping. When the vehicle stops running, during a traffic signal or when parking at the road side, the battery can be recharged by the ICE through the electric generator [25].

Hybrid electric vehicle systems 17

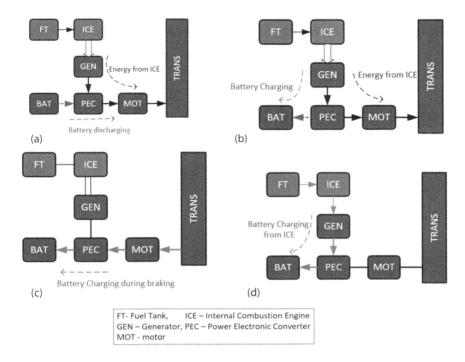

Figure 1.13 (a) Mode 1; (b) Mode 2; (c) Mode 3; (d) Mode 4.

1.5.2.2 *Power flow mode in parallel hybrid*

There are four modes of operation involved in the power flow control of a parallel hybrid system.

Mode 1 (at starting): Figure 1.14a shows the power flow mode of a parallel hybrid vehicle at starting. During this mode of power flow, the necessary energy is supplied by ICE and the electric motor to drive the vehicle wheel. The percentage of power shared by the ICE and the electric drive system is 80–20%.

Mode 2 (at normal driving): The power flow mode of a parallel hybrid vehicle at normal driving is shown in Figure 1.14b When the vehicle is under normal driving, the ICE only supplies the required power to the traction wheel, and the energy from the battery to the electric motor is in the off position.

Mode 3 (at braking): The parallel hybrid vehicle during braking mode is shown in Figure 1.14c When the brake is applied to the vehicle, the kinetic energy of the wheel transfers the motor operation into generator mode. The electricity generated by the motor during this operation is

18 *Power electronics for electric vehicles*

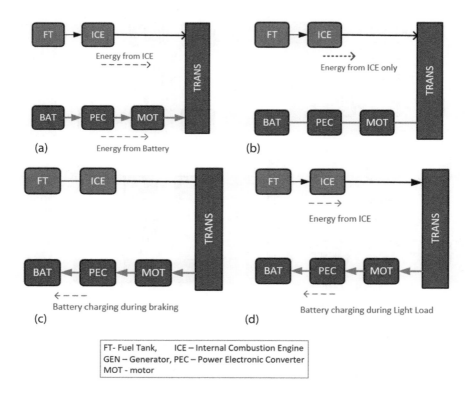

Figure 1.14 (a) Mode 1; (b) Mode 2; (c) Mode 3; (d) Mode 4.

supplied to the battery for charging [26]. The power electronic converter controls the power flow operation between the motor and the battery.

Mode 4 (at light load): The power flow mode of a parallel hybrid vehicle during light load conditions is shown in Figure 1.14d During this mode, the ICE supplies the traction power to the wheel and also charges the battery via the electric motor.

1.5.2.3 Power flow mode in series-parallel hybrid

There are five modes of operation involved in the power flow control of a series HEV system [3].

Mode 1 (at starting): Figure 1.15a shows the power flow mode of a series-parallel hybrid vehicle during starting. During this mode of operation, the rechargeable battery supplies the necessary power to drive the traction wheel, but the ICE remains in the off position.

Hybrid electric vehicle systems 19

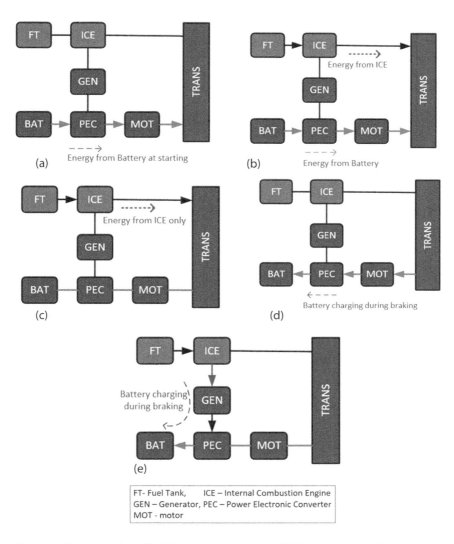

Figure 1.15 (a) Mode 1; (b) Mode 2; (c) Mode 3; (d) Mode 4; (e) Mode 5.

Mode 2 (at high-speed driving): The power flow mode of a series-parallel hybrid vehicle during high-speed driving is shown in Figure 1.15b When the vehicle needs high-speed driving, both the ICE and the electrical motor share the necessary drive power to the vehicle wheel.

Mode 3 (at normal driving): Figure 1.15 shows the power flow mode of a series-parallel vehicle under normal driving. During normal driving, the ICE only provides adequate drive power to the vehicle wheel; however, the electric motor remains in off mode [27].

20 Power electronics for electric vehicles

Mode 4 (at braking): The power flow mode of a series-parallel hybrid vehicle during braking is shown in Figure 1.15d During this mode of operation, the electric motor acts as a generator to supply the electrical energy to charge the battery. The power electronic controller helps to manage the power flow between the battery and the motor.

Mode 5 (battery charging at driving and standstill): Figure 1.15e shows the battery charging mode of a series-parallel hybrid vehicle at driving and standstill conditions. During normal driving, the ICE delivers the required power to drive the wheel and also recharge the battery. However, the ICE can also deliver the power to the battery through the electric generator at standstill conditions.

1.5.2.4 Power flow mode in complex hybrid

There are six modes of operation involved in the power flow operation of a complex hybrid vehicle.

Mode 1 (at Starting): Figure 1.16a shows the power flow mode of a complex hybrid vehicle at starting. During this mode of operation, the rechargeable battery will supply the necessary power to drive the traction wheel, but the ICE remains off during this mode of operation [3, 15].

Mode 2 (at high-speed driving): The power flow mode of a complex hybrid vehicle at high-speed driving is shown in Figure 1.16b. During this mode of operation, both the ICE and the front wheel electrical motor share the necessary power to the front wheel, and the second electric motor delivers the traction drive power to the rear wheel [3].

Mode 3 (at normal driving): Figure 1.16c shows the power flow mode operation of a complex hybrid vehicle at normal driving. When the vehicle is under normal driving mode, the ICE provides the necessary traction power to drive the front wheel and operate the first electric motor, run as a generator to recharge the battery [3].

Mode 4 (at light load driving): The power flow mode of a complex hybrid electric vehicle at light load driving is shown in Figure 1.16d. In this mode of operation, the required traction power to the front wheel is supplied by the first electric motor. The second electric motor and the internal combustion engines are off during this mode of operation.

Mode 5 (at braking): Figure 1.16e shows the power flow mode of a hybrid vehicle at braking. During this mode of operation, both the front and the rear wheel electric motors act as generators to simultaneously recharge the battery.

Mode 6 (at axle balancing): The power flow mode of a complex hybrid at axle balancing is shown in Figure 1.16f. The complex HEV has a unique feature of axle balancing. If any of the wheels slip (front or rear wheel), the corresponding electric motor (front or rear electric motor) runs as an electric generator by ICE power to charge the battery.

Hybrid electric vehicle systems 21

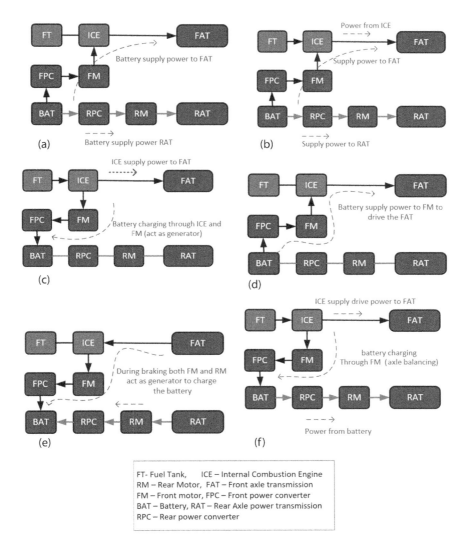

Figure 1.16 (a) Mode 1; (b) Mode 2; (c) Mode 3; (d) Mode 4; (e) Mode 5; (f) Mode 6.

1.5.3 Types of hybrid vehicle based on level or degree of hybridization

HEVs can be classified based on the proportion of driving power provided to the vehicle by the ICE and the electric drive system [28]. In some cases, the electric motor provides the dominant proportion, and in others, the ICE supplies the dominant proportion of drive power to the vehicle. The following are the various types of EV based on the level or degree of hybridization.

22 *Power electronics for electric vehicles*

1.5.3.1 *Mild hybrid*

Mild hybrid vehicles are basically the conventional vehicle types with larger starter motors, permitting the ICE to be off when the vehicle is driving under braking, coasting, stopping, and restarting. The high power electric motor is used to rotate the ICE to some driving speeds before allowing the fuel at lower speed whenever the ICE needs to be restarted. This concept is different from the HEV system because the hybrid system uses regenerative braking to recapture the energy. In a mild hybrid system, the larger motor is not acting like the motor used in an HEV system. Because of this, the fuel economy is lower than for a real HEV. Additional accessories need to be used for the air-cooling/conditioning system, which are usually run by the ICE. The lubrication system is the least effective immediately after the engine start every time. Furthermore, the continuous starting and stopping of this type of vehicle leads to reducing the lifetime of the ICE and thus, decreases the engine efficiency.

1.5.3.2 *Medium hybrid*

The medium hybrid electric system has a starter motor to boost the torque in the vehicle, which is coupled in parallel with the large powertrain. The electric mode of operation is possible only for a short period of time. Furthermore, the vehicle consumes less electric power, thus reducing the size of the battery. The starter motor is connected between the ICE and the mechanical transmission system. This motor runs when the ICE needs to be turned over and also when the vehicle needs additional power, which is run by gas. Furthermore, the electric motor can be used to restart the ICE, whereas the battery is used to supply the power to other vehicle accessories.

1.5.3.3 *Strong hybrid*

A strong hybrid is sometimes also called a full hybrid vehicle [29]. This vehicle can be run only on ICE, or run by an electric motor only, or run by the combination of both. When the vehicle is run only by battery-operated mode, it needs a large, high–power capacity rechargeable battery. These types of vehicle have the facility to split the drive power path between the electrical and mechanical power with greater flexibility.

1.5.3.4 *Plug-in hybrid*

The plug-in hybrid can also be called the plug-in hybrid electric vehicle (PHEV). This type of vehicle has a large, high-capacity rechargeable battery, which is able to store charge by connecting the battery to the external supply terminal with the help of a plug. The PHEV can be run by a battery supply system and also by ICE. For a short range of distance, the PHEV is independent of gasoline, and for long-distance travel, the vehicle

Hybrid electric vehicle systems 23

is dependent on both. For the serial type of plug-in hybrid, the ICE aids by providing the electrical drive power through an electrical generator if the vehicle is driving for a long distance. The PHEV can be manufactured by a multi-fuel system with the battery enhanced by gasoline or hydrogen. Compared with other HEV systems, the PHEV has a much larger all-electric range and reduces the "range anxiety" linked with other types of all-electric vehicles [30], because in a PHEV system, the ICE operates as a backup and supplies power when the rechargeable batteries are exhausted.

1.5.4 Types of hybrid vehicle based on nature of the power source

Depending on the nature of the drive power sources used, HEV systems can be classified into a few types.

1.5.4.1 Electric-ICE hybrid

The electric-ICE hybrid designs can be distinguished based on how the portions of electric and ICE powertrains are arranged, at what instant each portion of the system is in operation, and the percentage of drive power supplied by each component of the system. The ICE or battery can be turned on or off depending upon the type of operation in order to save energy.

1.5.4.2 Fuel cell–based hybrid

Fuel cell vehicles have a series hybrid configuration. A battery or super-capacitors are fitted in the vehicle to deliver the traction power during peak acceleration [31]. This decreases the size and power constraints on the fuel cell. In the fuel cell hybrid system, the fuel cell acts as the prime energy source, and a rechargeable battery or super-capacitors act as an energy storage system to ensure the fuel cell hybrid operation.

1.6 Conclusion

Increasing carbon emissions and decreasing fossil fuel energy sources are forcing vehicle manufacturers to develop transportation systems with numerous technologies. Among many other technologies in transportation systems, hybrid electric systems offer many advantages compared with other types of vehicle system: lower fuel consumption, less noise, smaller engine size, and longer operating life. This chapter presented an introduction to HEV systems with the historical development of all-electric vehicle systems, their electric vehicle technologies, configurations, and comparison. The various HEV drive configurations, such as series hybrid, parallel hybrid, series-parallel hybrid, and complex hybrid electric vehicle systems, have been discussed with the help of their operation modes. The energy flow from the ICE and the charging

24 *Power electronics for electric vehicles*

and discharging modes of the battery in the HEV, along with the operation of the electrical generator, motor, and power electronic controller, have been explored using power flow modes of operation. HEV systems with degrees of hybridization have been discussed based on how the ICE and the electric supply system share the energy in each mode of operation. The HEV system has more advantages compared with other vehicle technologies. Although it has a few disadvantages, the HEV is gaining recognition in the vehicle market because of its greater fuel economy and superior vehicle performance.

References

1. Haomiao Wang, Weidong Yang, Yingshu Chen, & Yun Wang. (2018) 'Overview of hybrid electric vehicle trend', AIP Publishing.
2. Zoran Stevic. (2012)'New generation of electric vehicles', InTech.
3. NPTEL. (2013) 'Introduction to hybrid and electric vehicle', Lecture Notes.
4. Interestingengineering.com. (2018). 'A brief history and evolution of electric cars'. Available at: https://interestingengineering.com/a-brief-history-and-evolution-of-electric-cars.
5. Zachary Shahan. (2015)'Electric car history'. Available at: https://cleantechnica.com/2015/04/26/electric-car-history/.
6. Aaron Smith, & Darren Moss. (2013) 'A brief history of hybrid and electric vehicles'. Available at: https://www.autocar.co.uk/car-news/frankfurt-motor-show/brief-history-hybrid-and-electric-vehicles-picture-special.
7. C. Chan. (2007) 'The state of the art of electric, hybrid, and fuel cell vehicles', *Proceedings of the IEEE*, 95(4), pp. 704–718.
8. Mehrdad Ehsani, Fei-Yue Wang, & Gary L. Brosch. (2013) *Transportation Technologies for Sustainability*. Springer.
9. A. Veziroglu, & R. Macário. (2013) 'Fuel cell vehicles (FCVs): State-of-the-art with economic and environmental concerns', in *Advances in Battery Technologies for Electric Vehicles, Handbook of Membrane Reactors: Reactor Types and Industrial Applications*, edited by Angelo Basile, Elsevier.
10. LalitKumar ShailendraJain. (2018) 'Fundamentals of power electronics controlled electric propulsion', in *Power Electronic Handbook*, pp. 1023–1065, Elsevier.
11. Pratama Mahadika, & Aries Subiantoro. (2019) 'Design of optimal controller for parallel hybrid electric vehicle based on shortest path algorithm', *International Seminar on Research of Information Technology and Intelligent Systems*, pp. 118–123.
12. Freedom National Auto Insurance Industry. (2019) *Advantages and Disadvantages of Electric Cars*. Freedom National.
13. U.S. Department of Energy. (2019) *Energy Efficiency & Renewable Energy*. Department of Energy.
14. Ohammad Kebriaei, Abolfazl Halvaei Niasar, & Behzad Asaei. (2015) 'Hybrid electric vehicles: An overview', *International Conference on Connected Vehicles and Expo (ICCVE), (10.1109/ICCVE)*, pp. 299–305.
15. C. C. Chan, & Y. S. Wong. (2004) 'Electric vehicles charge forward', *IEEE Power and Energy Magazine*, 2(6), pp. 24–33.

16. Krishna Veer Singh, Hari Om Bansal, & Dheerendra Singh. (2019) 'A comprehensive review on hybrid electric vehicles: Architectures and components', *Journal of Modern Transportation*, 77, pp. 77–107.
17. Fuad Un-Noor, Sanjeevikumar Padmanaban, Lucian Mihet-Popa, Mohammad Mollah, & Eklas Hossain. (2017) 'A comprehensive study of key electric vehicle (EV) components, technologies, challenges, impacts, and future direction of development', *MDPI, Open Access Journal*, 10(8), pp. 1–84.
18. Krishna Veer Singh, Hari Om Bansal, & Dheerendra Singh. (2019) 'A comprehensive review on hybrid electric vehicles: Architectures and components', *Journal of Modern Transportation*, 27, pp. 77–107.
19. C. C. Chan, A. Bouscayrol, & K. Chen. (2010) 'Electric, hybrid, and fuel-cell vehicles: Architectures and modelling', *IEEE Transactions on Vehicular Technology*, 59(2), pp. 589–598.
20. M. Ehsani. (2014) 'Conventional fuel/hybrid electric vehicles', in *Alternative Fuels and Advanced Vehicle Technologies for Improved Environmental Performance Alternative Fuels and Advanced Vehicle Technologies for Improved Environmental Performance*. Woodhead Publishing, pp. 632–654, edited by M. Ehsani.
21. M. Ehsani, Y. Gao, & J. M. Miller. (2007) 'Hybrid electric vehicles: Architecture and motor drives', *Proceedings of the IEEE*, 95(4), pp. 719–728. doi: 10.1109/JPROC.2007.892492.
22. Mehrdad Ehsani, Yimin Gao, & Ali Emadi. (2009) *Modern Electric, Hybrid Electric, and Fuel Cell Vehicles: Fundamentals, Theory, and Design*, 2nd edn. CRC Press.
23. F. Hu, & Z. Zhao. (2010) 'Optimization of control parameters in parallel hybrid electric vehicles using a hybrid genetic algorithm', *2010 IEEE Vehicle Power and Propulsion Conference*. Lille, pp. 1–6. doi: 10.1109/VPPC.2010.5729049.
24. Y. Cheng, R. Trigui, C. Espanet, A. Bouscayrol, & S. Cui. (2011) 'Specifications and design of a PM electric variable transmission for Toyota Prius II', *IEEE Transactions on Vehicular Technology*, 60(9), pp. 4106–4114. doi: 10.1109/TVT.2011.2155106.
25. M. Ehsani, N. Shidore, & Y. Gao. (2004) 'On board power management', *Power Electronics in Transportation (IEEE Cat. No.04TH8756)*, pp. 11–17, doi: 10.1109/PET.2004.1393783.
26. S. A. Zulkifli, N. Saad, S. Mohd, & A. R. A. Aziz. (2012) 'Split-parallel in-wheel-motor retrofit hybrid electric vehicle', *2012 IEEE International Power Engineering and Optimization Conference Melaka*, pp. 11–16. doi: 10.1109/PEOCO.2012.6230915.
27. Kyoungcheol Oh, Jeongmin Kim, Dongho Kim, Donghoon Choi, & Hyunsoo Kim. (2005) 'Optimal power distribution control for parallel hybrid electric vehicles', *IEEE International Conference on Vehicular Electronics and Safety*, pp. 79–85. doi: 10.1109/ICVES.2005.1563618.
28. O. D. Momoh, & M. O. Omoigui. (2009) 'An overview of hybrid electric vehicle technology', *2009 IEEE Vehicle Power and Propulsion Conference*, pp. 1286–1292.
29. V. T. Minh, & A. A. Rashid. (2012) 'Modeling and model predictive control for hybrid electric vehicles', *International Journal of Automotive Technology*, 13, pp. 477–485.

30. Mohammad Kebriaei, Abolfazl Halvaei Niasar, & Behzad Asaei. (2015) 'Hybrid electric vehicles: An overview', *2015 International Conference on Connected Vehicles and Expo (ICCVE)*, China, 2015, pp. 299–305.
31. Himadry Shekhar Das, Chee Wei Tan, & A. H. M. Yatim. (2017) 'Fuel cell hybrid electric vehicles: A review on power conditioning units and topologies', *Renewable and Sustainable Energy Reviews*, 76, pp. 268–291.

2 Onboard battery charging infrastructure for electrified transportation

Harish Karneddi and Deepak Ronanki

2.1 Introduction

Electric vehicles (EVs) are becoming tremendously attractive due to zero tailpipe emissions and eco-friendly natural transportation [1]. However, lack of charging infrastructure, capital cost, limited driving range, and the long idle period for charging hinder the fast adoption of EVs. The EV fleet numbered approximately 8.5 million globally in 2020 and is expected to increase by 58% by 2040. Federal and state governments are nurturing EVs into the transportation market with subsidies and tax exemptions [2]. Also, North American and European countries primarily provide subsidies and tax rebates to reduce the price margin between internal combustion engine vehicles (ICEVs) and EVs [3]. In recent times, the battery pack cost has fallen from $1200/kWh to $150/kWh and is expected to reduce further, by $70/kWh, by 2030, which is an excellent sign for a healthier and environmentally friendly transportation system through PEVs [4].

EVs are classified based on their architecture and energy sources into three types: hybrid EVs (HEVs), plug-in HEVs (PHEVs), and battery EVs (BEVs) or pure EVs. The HEV combines the advantage of both EV and ICEV, although batteries are charged from the ICE engine or during regenerative braking. The PHEV is similar to the HEV, but it can charge the battery from the additional charging equipment [5]. In pure EVs or BEVs, the propulsion torque is entirely generated by the battery alone, and the required energy to the battery is charged from the supply mains through the chargers.

The architecture/configuration of a BEV is illustrated in Figure 2.1. EVs involve several power conversion stages from the grid–battery–traction motor/auxiliary loads. The alternating current (AC) supply voltage from the grid through EV supply equipment (EVSE) is converted to direct current (DC), and that DC is configured to a compatible DC range to charge the battery. Isolation is provided to avoid leakage currents [6]. The battery energy is transferred to the traction motor or auxiliary load through the power converters based on their requirement [7]. EV chargers play a

DOI: 10.1201/9781003248484-2

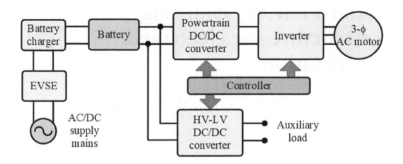

Figure 2.1 Architecture of the pure EV or BEV.

vital role in the EV's popularity, cost, and battery life. Over the past years, significant research has been carried out, and various charging topological configurations have been introduced to provide safe and efficient vehicle charging. Power electronic converters play a crucial role in enhancing the efficiency and power density of the EV charging infrastructure.

This chapter aims to give an overview of existing onboard charging topological structures for EVs. The configurations, standards, supply equipment of the EVs, and their charging protocols are discussed in detail. The main objective of this chapter is to provide a detailed overview of the state-of-art single-phase dedicated onboard chargers, including both two-stage and single-stage power conversion systems. A detailed case study on a single-phase 3.3 kW OBC is performed on the MATLAB/Simulink. Also, a detailed step-by-step design procedure is presented for the converter topology and its controller. Finally, the key performance of the boost PFC-based and interleaved boost PFC-based 3.3 kW OBC, such as power factor (PF), total harmonic distortion (THD), ripple current, and the output voltage ripple, is holistically evaluated by using case studies.

The chapter is organized as follows.

- The various aspects of EV charging infrastructure, standards, supply equipment, and their protocols are discussed in Section 2.2.
- Section 2.3 provides in-depth information on two-stage and single-stage dedicated onboard charging topologies.
- A status overview of single-phase onboard battery chargers is detailed in Section 2.4.
- A step-by-step design procedure of single-phase 3.3 kW conventional boost PFC-based OBC is presented in Section 2.5. Also, a comparative analysis of boost PFC and interleaved boost PFC-based OBC is illustrated.
- Finally, conclusive remarks are presented in Section 2.6.

2.2 EV charging infrastructure

EV chargers are classified into three types based on the power transfer modes: onboard chargers (OBCs), off-board chargers, and semi-onboard/off-board chargers. A detailed classification of the EV chargers is presented in Figure 2.2 [8].

OBCs are classified into dedicated and integrated OBCs; dedicated OBCs have a reliable converter inside the vehicle for charging purposes. Due to the size and weight constraints, the power rating of the dedicated OBCs is limited to lower values, resulting in a long idle period for charging [9]. On the other hand, several attempts have been made to propose integrated OBCs to enhance the power rating of the OBC without enlarging the size and weight. In such scenarios, charging will be performed during non-propulsion or vehicle idle cases. Integrated OBCs are designed with the help of a traction converter and/or traction motor to enhance the battery charger power ratings [10–12]. Alternatively, off-board chargers are external chargers designed with higher power ratings, which can mimic the functionality of fossil fuel refilling stations. Based on the local distribution among the chargers, off-board chargers are further classified into AC and DC distributed chargers [13, 14]. Semi-onboard/off-board chargers have half of the charger topology on the ground side and half on the vehicle side. Usually, wireless chargers are known as semi-onboard/off-board chargers. The transmitting side is equipped on the ground level side, and the receiving side is fitted on the vehicle side. Based on the coupling between transmitter and receiving side networks, wireless chargers are classified into inductive chargers (coupled with coils), capacitive chargers (coupled with capacitive plates), and hybrid chargers coupled with coils and metal plates [15–17].

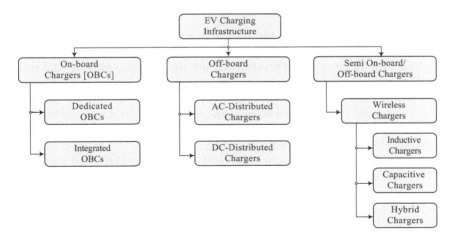

Figure 2.2 Classification of EV charging systems.

30 *Power electronics for electric vehicles*

2.2.1 EV standards

Governments have introduced various standards to standardize EVs and their charging systems. Some of the international standards, like International Electrotechnical Commission (IEC) 61581, IEC 62196, and IEC 62893, describe the general requirements of the charging infrastructures and all remaining parts, electromagnetic compatibility (EMC) requirements, charging cable requirements, and charging and discharging protocols, which are listed in Table 2.1 [6, 18–20].

2.2.2 Charging standards

Various charging standards are available to standardize battery chargers and are listed in Table 2.2. Charging solutions are realized with AC as well as DC power supplies. Based on the Society of Automotive Engineers (SAE) J1772, AC charging solutions can recharge up to 25 kW, and DC charging solutions can recharge up to 400 to 900 kW [21]. The second standard is "Charge-de-Move" (CHAdeMO), which the Japanese automakers and industry promote. These standards are promoted in three versions and can recharge the battery with 50 to 900 kW [22]. Due to the lack of charging infrastructure, the open charging system has faded through the years. The CharIn organization promoted a combined charging system (CCS), which can charge up to 200 to 400 kW. Chinese and Indian EV manufacturers, combined with elevated Guobiao standards (GB/T), can charge up to 900 kW and communicate with the CAN protocol. Finally, a fifth standard is the Tesla supercharger network promoted by Tesla. This can charge up to 350 kW, which is done through exclusive connections and proper cooling features [8].

2.2.3 EV supply equipment (EVSE)

EV chargers are connected to the vehicle from the supply mains (level-1 or level-2) or the off-board charging units (level-3) through different connectors based on the protocol. Therefore, connector type also plays a vital role in EV charging. Various connector configurations, also known as EVSE, are provided to the vehicle depending on the power rating of the charger and battery capacity. Based on the form of the power transfer, these connectors are classified into AC connectors and DC connectors. Some commercially available AC and DC connectors are mentioned in Table 2.3 [8, 13]. Among those connectors, some are compatible with both AC and DC, and some are suitable for single-phase and three-phase systems.

The circuit configuration of the EVSE is shown in Figure 2.3, consisting of five pins, i.e., line, neutral, earth, and two pilot pins. Based on the communication signal from the microcontroller unit (MCU), the input supply is connected to the vehicle [23]. According to the requirement, MCU controls the relays R_1 and R_2 in the conduction path. The proximity pilot line (PP) and

Table 2.1 International EV Standards

Standardization aspect	International standard	Part number	Description
EV conductive charging system	IEC 61581	1	General requirements for all the remaining parts.
		21-1	Electromagnetic compatibility (EMC) requirement for connectors.
		21-2	EMC compatibility for off-board charging infrastructure.
		3-1, 3-2, 3-3	Requirements for light EVs' (2-W, 3-W, cars) charging infrastructures.
		3-3 to 3-7	Requirements for light EVs' communication among various units, such as voltage converter unit, battery system.
	IEC TS 61581	23-1	Flash charging infrastructure for heavy EVs.
Connectors	IEC 62196	1, 2, 3, 3-1, 4, 6	Dimensional compatibility and interchangeability
	IEC 60309	1 to 5	requirements for AC/DC pins and connectors.
Cable safety measures	IEC 62752	–	Cable control and protection devices for mode-2 charging of EVs.
	IEC 62893	–	EV charging cables having a rating of 0.6/1 kV.
Switchgear protection and safety measures	IEC 61439	–	The standard for low-voltage switchgear and control gear infrastructure.
	IEC 60364	5-51 to 5-55	Includes earthing arrangements, isolation, switching, and control.
	IEC 60950		Low-voltage safety equipment (<600 V).
EMC compatibility	IEC 61851	1, 21-1, 21-2, 23 to 25	EMC and digital communication of EV chargers from the various charging stations (AC and DC).
	ISO 15118	–	Vehicle to grid communication.
	IEC 61000	–	Harmonic current limits,
	IEEE 519	–	voltage fluctuations, and power quality indices.
Charging protocol	IEC 63110	1 to 3	Charging and discharging protocol specifications and requirements.

Table 2.2 Charging Standards

SAE J1772 charging standards

AC charging standards

Charge method	Nominal input voltage [V]	Maximum current [A]	Input circuit breaker rating [A]	Power level [kW]
AC level 1	120	12	15	1.08
	120	16	20	1.44
AC level 2	208 to 240	16	20	3.3
	208 to 240	32	40	6.6
	208 to 240	≤ 80	Per NEC 635	≤ 14.4

DC charging standards

Charge method	Nominal input voltage [V]	Maximum current [A]	Power level [kW]
DC Level 1	200–450	≤ 80	≤ 36
DC Level 2	200–450	≤ 200	≤ 90
DC Level 3	200–600	≤ 400	≤ 240

CHAdeMO standards

Charging standard	Power rating [kW]	Output voltage [V]
CHAdeMO 1.0	50	50–1000
CHAdeMO 2.0	100 to 400	50–1000
CHAdeMO 3.0	~ 900	50–1000

(*Continued*)

Table 2.2 (Continued) Charging Standards

CCS standards

Charging standard	Power rating [kW]	Output voltage [V]	Region
CCS-1	350	200–1000	United States, South Korea
CCS-2	350	200–1000	Europe, Australia

GB/T

Charging standard	Power rating [kW]	Output voltage [V]	Input supply
GB/T 20234-2	7/12.8	200–750	1-ϕ/3-ϕ
GB/T	237.5	250–950	DC

Tesla supercharging network

Power rating	Output voltage	Maximum current	Input supply
250–350 kW	300–480 V	800 A	DC

Table 2.3 EVSE Connector Specifications

DC connectors				
Connector type	GB/T 20234 DC	CHAdeMO	CCS Combo 1	CCS Combo 2
Connector configuration				
No. of pins	9 2-DC lines2-Signal lines2-Connection checks2-Auxiliary circuit supply1-Protective earth pin	9 2-DC lines2-Signal lines2-Start/Stop pins1-Connection check1-Charging enable/disable1-Protective earth pin	7 2-AC lines2-DC lines2-Signal lines1-Protective earth pin	9 3-AC lines1-Neutral pin2-DC lines2-Signal lines1-Protective earth pin
Voltage rating	120 or 240 V	400 V	380 V	480 V
Power rating	187.5 kW	200 kW	75 kW	200 kW
Compatibility	DC	DC	DC and AC	DC and AC

(*Continued*)

Table 2.3 (Continued) EVSE Connector Specifications

AC connectors

Connector type	SAE J1772 Type 1	IEC 62196-2 Type 2	GB/T 20234 AC	Tesla Supercharger
Connector configuration				
No. of pins	5	7	7	5
	2-AC lines2-Signal lines1-Protective earth pin	3-AC lines1-Neutral pin2-Signal lines1-Protective earth pin	3-AC lines1-Neutral pin2-Signal lines1-Protective earth pin	2-AC lines2-Signal lines1-Protective earth pin
Voltage rating	750 V	500 V	600 V	1000 V
Power rating	7.68 kW	12.8 kW	12.16 kW	140 kW
Compatibility	AC	AC	DC and AC	DC and 1-ϕ AC

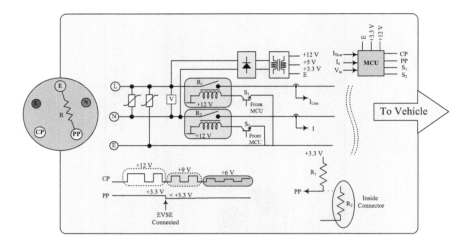

Figure 2.3 Illustration of EVSE circuit configuration.

control pilot line (CP) generate the signals to control the power flow. The PP pilot line has 3.3 V when the EVSE is not connected to the vehicle. Once the EVSE is connected to the vehicle, external resistance R_2 is added in series with R_1 and reduces the voltage at PP. The CP line has a voltage of −12 V at its terminals. Once the EVSE is ready, the voltage is changed from −12 V to +12 V. Once the EVSE is ready, the voltage is reduced to +9 V, and the PWM signal starts pulsating between +9 and −12 V at 1 kHz. The duty cycle of the PWM signal conveys the power availability at charger terminals. Finally, CP voltage is reduced to +6 V, and the EV starts charging. After completion of charging, similarly, the CP signal reaches +12 V. Once the EVSE is disconnected from the vehicle, the voltage level at the PP increases, and the EVSE understands it is disconnected from the vehicle and stops charging.

2.3 Dedicated OBCs

The dedicated OBCs are classified into various categories based on input power supply, power flow direction, conversion stages, and isolation.

- Based on input power supply, OBCs are categorized into:
  - Single-phase OBCs
  - Three-phase OBCs
- Based on the power flow direction between the supply mains and the vehicle, OBCs are classified into:
  - Unidirectional OBCs (grid to vehicle)
  - Bidirectional OBCs (grid to vehicle (G2V)/vehicle to grid (V2G))

- Based on the number of conversion stages involved between the grid and the vehicle, OBCs are classified into:
  - Two-stage OBC (AC–DC–DC)
  - Single-stage OBC (AC–DC)
- Based on the isolation, OBCs are classified into:
  - Non-isolated OBCs
  - Isolated OBCs

Due to the switching operations, power electronic converters will inject the current harmonics into the grid and draw the reactive power. Therefore, the onboard chargers are designed to maintain sinusoidal input current with a unity PF and the ripple-free DC voltage at the battery terminals. In two-stage onboard chargers, the front-end PFC converter maintains the sinusoidal current with unity power factor and maintains ripple-free DC link voltage [24–26]. The DC–DC interface supports the ripple-free DC voltage with galvanic isolation at the battery terminals. In the case of single-stage onboard battery chargers, all the aforementioned requirements are fulfilled by the one conversion stage only. This reduces the number of conversion stages and costs and improves efficiency [27].

2.3.1 Two-stage OBCs

Two-stage OBCs [Figure 2.4] consist of two conversion stages from the grid to the battery terminals, i.e., the PFC circuit and the battery interface converter. The PFC stage consists of the rectification and DC–DC conversion stages. The battery interfacing stage facilitates galvanic isolation with the DC–DC conversion stage [27].

2.3.2 PFC stage

The PFC stage contains the rectification stage followed by the DC–DC stage to perform the various operations listed here:

- Maintains unity PF (UPF) at the grid side
- Minimizes the input current total harmonic distortion (THD)

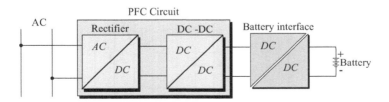

Figure 2.4 Block diagram of two-stage OBC.

- Minimizes input current ripple
- Maintains the required DC voltage with minimal voltage ripple at the DC link

PFC converter controls such that it performs well to fulfil the aforementioned requirements. Based on the charging current, these converters are operated in continuous conduction mode (CCM) or discontinuous conduction mode (DCM). The design of controllers for PFC converters is different when they are operating at either CCM or DCM. The DCM PFC converter uses either constant or variable frequency control operation. Direct current mode control (hysteresis current control, peak current mode control, and average current mode control) and indirect current control methods are available for CCM-based PFC converters [28–30]. Among all controls, average current mode control has more advantages, such as minimal THD and electromagnetic interference (EMI) emission and is suitable for both CCM and DCM operations.

Based on the converter configuration structures, PFC converters are classified into:

- Bridge-type PFC topologies
- Bridgeless PFC topologies

2.3.2.1 Bridge PFC converters

Bridge PFC converters consist of an uncontrolled front-end rectifier followed by the DC–DC converter. An uncontrolled rectifier, typically a diode bridge rectifier, converts the input AC to DC. The DC–DC converter controls the input PF, ripple current, and THD and maintains the stiff DC voltage profile at the DC link. Due to its simple design and control, a conventional boost converter is used as a DC–DC conversion stage. It consists of an input side inductor resulting in continuous input current with minimized ripple [24, 31]. Figure 2.5a illustrates the circuit configuration of the

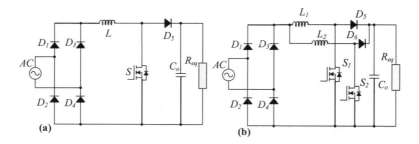

Figure 2.5. Bridge PFC converter: (a) boost PFC and (b) interleaved boost PFC.

boost PFC converter. Two or more boost PFC converters are connected in parallel and are operated out of phase to reduce the input current ripple further, which is termed an interleaved boost PFC converter [Figure 2.5b.

The operating modes of the bridge boost PFC converter are illustrated in Figure 2.6. The bridge rectifier provides a pulsating DC at the output terminals of the rectifier. Therefore, boost converter operation remains the same for positive or negative half-cycles. Figure2.6a–d shows all possible conduction modes. Figure 2.6a and b represent the circuit configuration for the $D*T_s$ period and the $(1-D)*T_s$ period for the positive half-cycle of input, respectively.

Similarly, conduction paths for the negative half-cycle are illustrated in Figure 2.6c and d. During the negative half-cycle, the diodes D_2 and D_3 come into conduction. The negative half-cycle input is applied as a positive half-cycle at the input terminals of the boost converter. Therefore, the operation of the DC–DC converter remains the same in both the half-cycles.

The bridge PFC converters consist of a minimum of three switches in all the conduction paths [Figure 2.6], thereby causing more conduction losses. Therefore, the bridge converters are preferable for low- and medium-power applications. The boost PFC converter is simple in design and to control and hence suitable for low-power applications due to the greater input current ripple. The interleaved boost PFC converter is ideal for medium-power

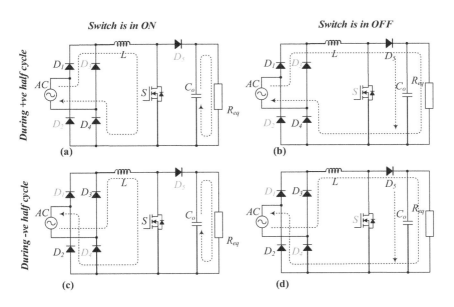

Figure 2.6 Conduction modes of the boost PFC circuit: (a) positive half-cycle-DT_s period; (b) positive half-cycle $(1-D)T_s$ period; (c) negative half-cycle DT_s period; and (d) negative half-cycle $(1-D)T_s$ period.

40 Power electronics for electric vehicles

Table 2.4 Comparison of the Bridge PFC Converters

Parameters	Conventional Boost PFC	Interleaved Boost PFC
No. of controlled switches	1	2
No. of diodes	5	6
No. of inductors – size	1 – L	2 – L/2
Inductor location	DC side	DC side
Suitable power rating	~1 kW	~3 kW
Input current ripple	More	Less
Output voltage ripple	More	Less
Efficiency	Low	High
Cost	Low	High

applications. A detailed comparison of bridge PFC topologies is presented in Table 2.4 [26].

2.3.2.2 Bridgeless PFC

Bridgeless PFC converters are developed to avoid the losses associated with the front-end uncontrolled bridge rectifier stage in the PFC converter. The most popular bridgeless PFC converters are illustrated in Figure 2.7. Figure 2.7a depicts the bridgeless boost PFC converter, consisting of two controlled switches with anti-parallel diodes and two diodes [32]. Both the switches are operated with the common gate pulse. Both the switches S_1 and S_2 are closed during the $D*T_s$ period (irrespective of input) and charge the boost inductor. During the positive half-cycle, the charged inductor is discharged through D_1 and the body diode of the S_2.

Similarly, in the negative half-cycle of AC input, the boost inductor and the source combined deliver energy through D_2 and the body diode of S_1. Only two devices are in the conduction path; therefore, the losses associated with the bridgeless boost PFC converter are reduced compared with the bridge PFC converters [33]. Two or more bridgeless boost PFC converters

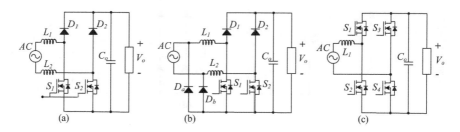

Figure 2.7 Bridgeless PFC converters: (a) bridgeless boost PFC, (b) semi-bridgeless, and (c) totem-pole PFC.

Onboard battery charging infrastructure 41

are cascaded together to form an interleaved bridgeless boost PFC converter to reduce the input current ripple further and are operated out of phase.

In the bridgeless boost PFC converter, the output DC bus is floating, leading to sizeable common-mode noise. Semi-bridgeless PFC converters are proposed to reduce the common-mode noise by connecting the output bus to the input with two line frequency operated diodes [Figure 2.7b. These line frequency diodes (D_a and D_b) come into conduction during CCM. Instead of line frequency diodes, capacitors can also realize the same operation [34]. Line frequency diodes bypass one inductor from the conduction path, realizing the same ripple current as bridgeless boost PFC. Therefore, this converter requires double the inductor size compared with a bridgeless boost PFC to maintain the same ripple. The two converters described so far are most suitable for unidirectional power flow due to unidirectional devices.

Figure 2.7c depicts the totem-pole PFC converter employing controlled switches. Two switches (S_4-positive half-cycle and S_3-negative half-cycle) are operated with line frequency, and the remaining two are operated in a complementary manner with the switching frequency (f_s). During the positive half-cycle, S_2 and in the negative half-cycle, S_1 is operated as a boost switch, i.e., turned on for the $D*T_s$ period. The switching losses and conduction losses associated with the totem-pole PFC converter are lower due to fewer devices in the conduction path, as well as half of the switches being operated at line frequency. Therefore, totem-pole PFC is preferable for high-power and bidirectional power flow applications [35]. The detailed comparisons of the bridgeless boost PFC converters are presented in Table 2.5.

2.3.2.3 DC–DC battery interface converter

This conversion stage facilitates galvanic isolation between the grid/source and the vehicle to avoid touch currents. It also provides the desired ripple-free

Table 2.5 Comparison of Bridgeless PFC Converters

Parameters	Bridgeless boost PFC	Semi-bridgeless boost PFC	Totem-pole PFC
No. of controlled switches	2	2	4
No. of diodes	2	4	0
No. of inductors – size	2 – L/2	2 – L	1 – L
Suitable power rating	<2 kW	<4 kW	>6 kW
Output voltage ripple	High	Medium	Less
Efficiency	Low	Medium	High
Cost	Low	High	Medium

DC voltage at the input terminals of the battery with high efficiency and lower EMI emission. Output ripple can be reduced with an increase in filter size or by an increase in the converter's switching frequency. An increase in filter size increases the size and cost of the charger; therefore, it's not a feasible solution. An increase in switching frequency decreases the filter size. However, it increases the switching losses in the conventional isolated DC–DC converters and thereby decreases the charger efficiency. Therefore, resonant and soft-switching circuits are widely adopted to operate at higher frequencies [27].

Several resonant converter topologies that can be utilized for the DC–DC battery interface are shown in Figure 2.8. The resonant converter topologies consist of a front-end square wave inverter followed by the resonant circuit, isolated high-frequency transformer, and rectifier. Figure 2.8a depicts the series resonant DC–DC converter. The inverter is realized with the half-bridge square wave inverter. The resonant circuit comprises the series-connected L and C elements [36]. Series resonant tank impedance and the load impedance act as a voltage divider, and the voltage across the load can vary by a change in the resonant frequency. The resonant tank and load impedance are connected in series with the input supply and result the load voltage are always less than the input voltage. Therefore, the series resonant DC–DC converter gain is always less than unity. Furthermore, it is difficult to control the voltage across the load during light-loaded conditions [27].

Figure 2.8b shows the parallel resonant DC–DC converter. The resonant converter is formed by the L and C_1, which are connected in parallel to the load. Therefore, the gain of the parallel resonant converter is more significant than the unity. A parallel resonant converter avoids the difficulty associated with a series resonant converter, i.e., light load voltage control. However, the operating range of the parallel resonant converter is much

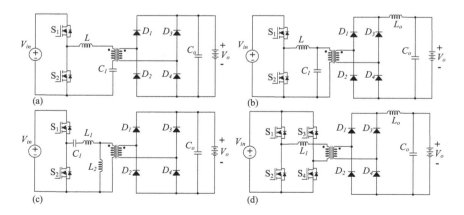

Figure 2.8 DC–DC interface converters: (a) series resonant, (b) parallel resonant, (c) LLC resonant, and (d) phase-shifted full-bridge converters.

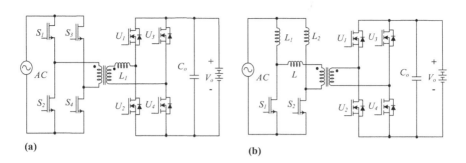

Figure 2.9 Single-stage onboard chargers: (a) matrix converter-based single-stage OBC [41] and (b) single-stage bidirectional OBC [43].

smaller. The load side inductor L_o is connected for impedance matching. Combining series and parallel resonant tanks to merge the advantages of both the converters forms a series-parallel resonant converter. The resonant network is formed with three filter elements in various forms, and one of the configurations, i.e., the LLC resonant converter, is presented in Figure 2.9c [37]. Compared with the previously described converters, the LLC resonant converter gives a better performance from the perspective of efficiency, switching loss, voltage regulation, and circulating currents [38].

On the other hand, the PSFB converter illustrated in Figure 2.9d consists of a front-end full-bridge square wave inverter followed by a high-frequency transformer (HFT) and a rectifier. The rectifier can model uncontrolled or controlled switches depending on the power flow (G2V or V2G) requirement [39]. The gain of the PSFB converter is also less than unity, and the gain of the whole system can increase with an increased transformation ratio. A secondary side filter inductor matches the load impedance, similarly to the parallel resonant converter [40]. The critical features of various DC–DC resonant converter topologies are summarized in Table 2.6.

2.3.3 Single-stage OBC

The single-stage OBC consists of only one conversion stage from the grid to the battery. This conversion stage serves all the requirements of the OBC. A single converter facilitates all the functionalities provided by the PFC stage and the DC–DC battery interface stage, i.e., maintains ripple-free current with minimal THD and unity PF at the input. Also, it offers the required DC voltage at the output terminals and galvanic isolation. Single-stage OBCs can increase the power density of the charger by avoiding the DC link capacitor, and their topologies are shown in Figure 2.9.

Figure 2.9a shows that the single-stage matrix converter-based single-stage OBC comprises a matrix converter followed by the HFT and dual active bridge (DAB). The matrix converter converts line frequency AC to

Table 2.6 Comparison of DC–DC Interface Converter Topologies

Battery Interface	Series Resonant Converter	Parallel Resonant Converter	LLC Resonant Converter	PSFB Converter
No. of controlled switches	2	2	2	4
No. of uncontrolled switches	4	4	4	4
No. of filter elements	2	2	3	1
Suitable power	Low	Medium	Low and medium	High
Efficiency	Low	Medium	High	High
Operating frequency range	Wide	Wide	Limited	Wide
Switching stress	Low	High	Low	High
Features	• Simple in design • Minimum EMI emission • Poor voltage regulation • Gain <1	• Better voltage regulation • Suitable for higher current and low output voltages • More circulating currents	• Lower EMI emission • Lower cost	• Zero voltage switching is possible • High primary circulating currents • Expensive

high-frequency AC with the help of gallium nitride (GaN) devices, and the secondary side DAB controls the input PF and output voltage. Controlling this OBC is the same as the aforementioned single-stage topology control [41–42]. This OBC is suitable for bidirectional power flow, i.e., G2V and V2G power are possible. This charger's power density and cost are higher than for the charger mentioned earlier due to the absence of a DC link capacitor and bidirectional switches. In a single-stage bidirectional OBC, shown in Figure 2.9b, the secondary side is similar to a matrix converter-based OBC. However, the primary side configuration is reformed to reduce the number of switching components. The primary side converter is modelled as an interleaved current-fed half-bridge configuration with zero current switching (ZCS), and the secondary is similar to the previous topologies [43]. Due to the interleaved configuration, input current ripple is reduced, and the harmonic component also shifts to twice the previous. As a result, a reduction in filter size and an improvement in the power density of the charger are observed. Overall, the single-stage converters have various advantages, such as a minimum number of conversion stages, components, and maximum power density.

2.4 Technological overview of OBCs

Various OBCs have been developed by several industrial manufacturers such as Brusa, Chevy Volt, Tesla, BorgWarner, etc. Figure 2.10 illustrates the circuit configuration of the Delta-Q OBC configured with an interleaved boost converter as a PFC circuit and a PSFB converter as a DC–DC interface. Delta-Q technologies developed an OBC for 3.3 kW to operate at a universal AC input voltage to generate 250-430 V with an efficiency of 94% [31]. The Chevy second-gen Volt 3.3 kW OBC also has a similar configuration. The front-end PFC stage is realized with interleaved boost PFC. The PSFB is utilized for the DC–DC battery interface stage [43].

A summary of commercially available OBCs is listed in Table 2.7. Among them all, few chargers are compatible for single-phase as well as three-phase.

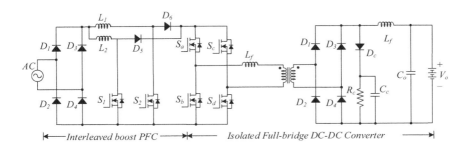

Figure 2.10 Circuit configuration of Delta-Q OBC.

46   *Power electronics for electric vehicles*

Table 2.7 Commercially Available OBCs

Onboard charger topology	Suitable input voltage [V]	Power rating of the charger [kW]	Output voltage [V]	Efficiency [%]
Second-gen Chevy Volt [43]	Universal AC	3.65	200 to 420	92
Hyundai vehicle to the grid [44]	Universal AC	7.2	350	>90
Tesla Model 3 [8]	1- or 3-phase	7.5/11.5	350	-
Multifunctional OBC [45]	110	2	250 to 430	96.1
Delta-Q technologies [31]	Universal AC	3.3	200 to 450	94
Current Ways [46]	Universal AC	6.6	250 to 450	>90
Overtech [47]	Universal AC	6.6	200 to 420	>90
Brusa [48]	200–250	6.5 to 20	200 to 450	>94
BorgWarner [49]	1- or 3-phase	1.8 to 22	400 to 650	–

2.5 Case study of 3.3 kW two-stage OBC

An OBC is designed for the specifications mentioned in Table 2.8 for a 3.3 kW second-gen Volt charger [43]. A universal input voltage, i.e., 85 to 264 V, is considered for the design of the front-end PFC stage. The DC link voltage is considered to be 400 V with 10 V_{p-p} ripple, and for these specifications, all the components are calculated. The holding time of the circuit is considered the period of one complete cycle. The output filter capacitor is calculated based on the holding time and the ripple voltage and is considered the more significant value. Similarly, PSFB converter component values are calculated per the specifications mentioned in Table 2.8.

Table 2.8 Specifications of a Designed 3.3 kW Onboard Charger

Front-End Converter		DC-DC Battery Interface	
Boost/Interleaved boost		PSFB	
Specifications	**Values**	**Specifications**	**Values**
Input Voltage	85 to 260 V	DC Link Voltage	400 V
Input Frequency	50 Hz ± 1%	Operating Frequency	100 kHz
Power	3.3 kW	Power	3.3 kW
Current Ripple	20%	Output Voltage	120 V
DC link Voltage	400 V	Voltage Ripple	10 V_{p-p}
Voltage Ripple	10 V_{p-p}	Current Ripple	20%
Switching Frequency	50 kHz		
t_{hold}	20 ms		

Onboard battery charging infrastructure 47

2.5.1 Modelling of the boost PFC converter

Modelling of the PFC converter is needed to calculate the component values for the desired specifications. The average model is employed to model the PFC converter for the controller design. Two states of operation are considered: during switch on (Figure 2.6a and c), i.e., $D*T_s$ period, and during switch off (Figure 2.6b and d), i.e. $(1-D)*T_s$ period.

Mode-1: State equations for the $D*T_s$ period are

$$L\frac{di_L}{dt} = V_{in} \tag{2.1}$$

$$C\frac{dv_c}{dt} = -\frac{v_c}{R} \tag{2.2}$$

Equations (2.1) and (2.2) can be written in the form of a state-space model, i.e.,

$$\begin{bmatrix} \dot{i}_L \\ \dot{v}_c \end{bmatrix} = \begin{bmatrix} 0 & 0 \\ 0 & \dfrac{-1}{RC} \end{bmatrix} \begin{bmatrix} i_L \\ v_c \end{bmatrix} + \begin{bmatrix} \dfrac{1}{L} \\ 0 \end{bmatrix} V_{in} \qquad i.e.\,\dot{x} = A_1 x + B_1 u \tag{2.3}$$

Mode-2: State equations for the $(1-D)*T_s$ period are

$$L\frac{di_L}{dt} = V_{in} - v_c \tag{2.4}$$

$$C\frac{dv_c}{dt} = i_L - \frac{v_c}{R} \tag{2.5}$$

Equations (2.4) and (2.5) can be written in the form of a state-space model, i.e.,

$$\begin{bmatrix} \dot{i}_L \\ \dot{v}_c \end{bmatrix} = \begin{bmatrix} 0 & \dfrac{-1}{L} \\ \dfrac{1}{C} & \dfrac{-1}{RC} \end{bmatrix} \begin{bmatrix} i_L \\ v_c \end{bmatrix} + \begin{bmatrix} \dfrac{1}{L} \\ 0 \end{bmatrix} V_{in} \qquad i.e.: \dot{x} = A_2 x + B_2 u \tag{2.6}$$

From both conduction mode state equations, the average model of the converter can be written as

$$A = d*A_1 + (1-d)*A_2$$

$$B = d*B_1 + (1-d)*B_2 \tag{2.7}$$

48 *Power electronics for electric vehicles*

Therefore, from the equations (2.3), (2.6), and (2.7), the total state-space model of the PFC converter can be modelled as

$$
\begin{bmatrix} \dot{i}_L \\ \dot{v}_c \end{bmatrix} = \begin{bmatrix} 0 & \dfrac{-(1-d)}{L} \\ \dfrac{(1-d)}{C} & \dfrac{-1}{RC} \end{bmatrix} \begin{bmatrix} i_L \\ v_c \end{bmatrix} + \begin{bmatrix} \dfrac{1}{L} \\ 0 \end{bmatrix} V_{in}
\tag{2.8}
$$

$$
A = \begin{bmatrix} 0 & \dfrac{-(1-d)}{L} \\ \dfrac{(1-d)}{C} & \dfrac{-1}{RC} \end{bmatrix}
\tag{2.9}
$$

$$
B = \begin{bmatrix} \dfrac{1}{L} \\ 0 \end{bmatrix}
\tag{2.10}
$$

2.5.2 *Component calculation*

Assume that the converter is lossless. This section presents the design equations of the PFC and PSFB converters. All component values are calculated for boost PFC-based and interleaved boost PFC-based OBCs and are shown in the following.

2.5.2.1 *Boost PFC design*

The modelling and design equations for the boost converter are presented in the following [50].

The peak current of the boost inductor is given by

$$
I_{pk\_PFC} = \frac{\sqrt{2}P_{in}}{V_{in\_min}}
\tag{2.11}
$$

Where P_{in} is the input power of the converter = 3.3 kW

V_{in_min} is the minimum input voltage = 85 V_{RMS}

$$
I_{pk\_PFC} = 54.90 \text{ A}
$$

The ripple current of the boost inductor is calculated as

$$
\Delta I_{l\_PFC} = 0.2 * I_{pk\_PFC}
\tag{2.12}
$$

$$
\Delta I_{l\_PFC} = 10.980 \text{ A}
$$

The duty cycle of the boost converter is calculated as

$$D_{PFC} = \frac{V_{o\_PFC} - \sqrt{2}V_{in\_min}}{V_{o\_PFC}} \tag{2.13}$$

Where V_o is the output voltage of the PFC converter = 400 V

$$D_{PFC} = \frac{400 - \sqrt{2}*85}{400} = 0.7$$

The inductance of the boost inductor is calculated as

$$L_{PFC} = \frac{\sqrt{2}V_{in\_min}*D_{PFC}}{f_{s\_PFC}*\Delta I_{l\_PFC}} \tag{2.14}$$

Where f_s is the switching frequency of the PFC converter = 50 kHz

$$L_{PFC} = \frac{\sqrt{2}*85*0.7}{50000*10.98} = 0.1532\,mH$$

The capacitance of the output capacitor is calculated as

$$C_o{}' = \frac{2*P_{in\_PFC}*t_{hold}}{V_{o\_PFC}{}^2 - V_{o\_min\_PFC}{}^2} \tag{2.15}$$

Where t_{hold} is the holding time of the PFC converter = 20 ms

$$C_o{}' = \frac{2*3300*0.02}{400^2 - 350^2} = 3.52\,mF$$

2.5.2.2 Phase-shifted full-bridge converter (PSFB)

The design equations of the PSFB converter are presented as follows:

The transformation ratio (n) of an high-frequency transformer (HFTF) is given by

$$\frac{N_s}{N_p} = \frac{V_s}{V_p} \tag{2.16}$$

Where N_s is the no. of turns on the secondary
N_p is the no. of turns on the primary
V_s is the secondary voltage of the transformer
V_p is the primary voltage of the transformer

50   *Power electronics for electric vehicles*

$$n = \frac{150}{400} = 0.375$$

The peak current through the PSFB inductor is calculated as

$$I_{pk\_PSFB} = \frac{P_o}{V_{o\_min}} = \frac{3300}{120} = 27.50\,A$$

The ripple inductor current of the PSFB converter is given as

$$\Delta I_{l\_PSFB} = 0.2 * I_{pk\_PSFB} = 5.5\,A$$

The duty cycle of the PSFB converter is calculated as

$$D_{PSFB} = \frac{V_o}{2 * V_{max}} \tag{2.17}$$

Where V_{max} maximum available voltage at secondary = V_s = 150 V

$$D_{PSFB} = \frac{120}{2 * 150} = 0.625$$

The voltage across the inductor is calculated as

$$V_{L\_PSFB} = V_{in\_PSFB} * \frac{D_{PSFB}}{f_{PSFB} * \Delta I_{l\_PSFB}} - V_o \tag{2.18}$$

$$V_{L\_PSFB} = 400 * \frac{0.625}{100k * 5.5} - 120 = 30\,V$$

$$V_{L\_PSFB} = L * \frac{di}{dt} \tag{2.19}$$

The inductance is calculated as

$$L_{PSFB} = V_{L\_PSFB} * \frac{D_{PSFB} * T_{s\_PSFB}}{\Delta I_{l\_PSFB}} = 34.0909\,\mu H \tag{2.20}$$

The charging current is given by

$$I_{C\_PSFB} = C_o * \frac{dV_o}{dt} \tag{2.21}$$

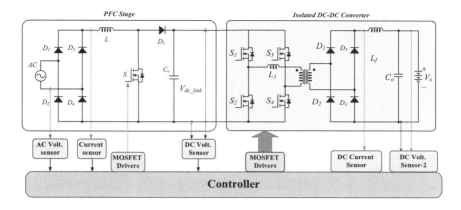

Figure 2.11 Architecture of conventional boost PFC-based OBC.

The capacitance of the output filter can be calculated as

$$C_o = \Delta I_{l\_PSFB} * \frac{D_{PSFB}}{f_{PSFB} * \Delta V_0} = 28.6458\,\mu F \qquad (2.22)$$

Both PFC converters and the DC–DC converters are modelled on the MATLAB/Simulink platform with the derived component values. The complete architecture of the OBC is illustrated in Figure 2.11. An individual controller controls the PFC and isolated DC–DC converters. The PFC controller provides the duty cycle for the PFC converter. The PFC converter maintains the unity PF at the source terminals, extracts the harmonic free current, and maintains the ripple-free voltage at the DC link. The outer voltage loop monitors DC link voltage and generates the reference current. The inner current loop controls the unity PF and the harmonic free current. The reference current from the previous stage is compared with the actual current and passed through the controller to generate the duty cycle. The duty cycle is compared with the 50 kHz sawtooth signal to generate the gate pulse and connected to the boost switch. The PSFB DC–DC converter is controlled with a dual loop controller, consisting of the inner current loop and the outer voltage loop. The inner loop controls the output current during the constant current (CC) charging mode, whereas the output voltage loop maintains the desired voltage during the constant voltage (CV) mode of charging.

2.5.3 Controller design

The complete control architectures for both stages are illustrated in Figure 2.12. Each controller consists of the inner current loop and the outer

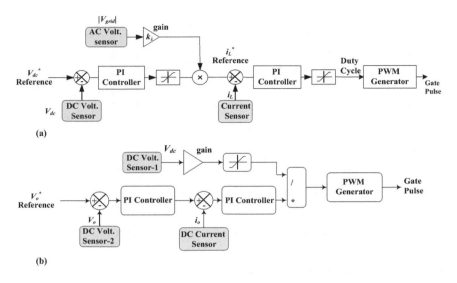

Figure 2.12 Control circuit architecture of two-stage OBC: (a) boost PFC converter and (b) isolated DC–DC converter.

voltage loop. The inner current loop is faster than the outer voltage loop to follow the sinusoidal reference current to maintain the unity PF. In general, the current loop bandwidth is relatively high; it's ten times lower than the switching frequency to filter the harmonics. The error voltage is the input for the outer voltage loop. However, error voltage consists of the second harmonic component due to the second harmonic ripple voltage in the output voltage, resulting in a third harmonic reference current generation for the inner current loop. Therefore, a voltage controller is designed with a lower crossover frequency (usually ten times the line frequency) to suppress the second harmonic voltage component in error voltage. A similar kind of control structure is used for the PSFB converter.

The interleaved boost PFC-based OBC is simulated with a similar control architecture on the MATLAB/Simulink platform. Two pulses are generated with two separate inner current loops for the interleaved boost PFC. The typical outer voltage loop generates a reference current, and this is compared with the actual inductor currents and passed through individual duty cycles d_1 and d_2.

The sawtooth waves used to generate switching pulses are out of phase with each other. The controller for the interleaved boost PFC converter is illustrated in Figure 2.13. The isolated PSFB converter is used as a DC–DC interface for both cases. The proportional-integral (PI) controller is used

Onboard battery charging infrastructure 53

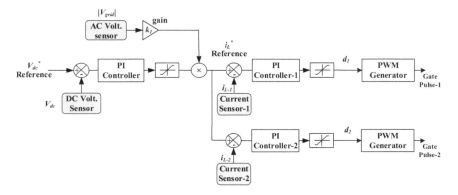

Figure 2.13 Controller architecture for interleaved boost PFC converter.

for all the controllers. Each controller's gain coefficients k_p and k_i are determined using a Bode plot.

The state-space model of the boost PFC converter (equations (2.8)–(2.10)) can be represented in a transfer function model as follows.

$$\frac{i_L(s)}{d(s)} = \frac{C * V_o * s + 2 * (1-D) * I_l}{(L*C)*s^2 + \frac{L}{R}*s + (1-D)^2} \quad \text{......inner current loop} \quad (2.23)$$

$$\frac{v_o(s)}{d(s)} = \frac{(L*I_l)*s - (1-D)*V_o}{(L*C)*s^2 + \frac{L}{R}*s + (1-D)^2} \quad \text{......outer voltage loop} \quad (2.24)$$

The inner current loop transfer function of the boost PFC converter is

$$\frac{i_L(s)}{d(s)} = \frac{1.408*s + 32.94282}{0.5393*10^{-6}*s^2 + 0.007427*s + 0.09} \quad (2.25)$$

The outer voltage loop transfer function of the boost PFC converter is

$$\frac{v_o(s)}{d(s)} = \frac{0.008411*s - 120}{0.5393*10^{-6}*s^2 + 0.007427*s + 0.09} \quad (2.26)$$

The Bode plot of the open-loop transfer function is plotted with the help of MATLAB sisotool. The k_p and k_i values are calculated. The bandwidth of the current loop is maintained at ~5000 Hz, and similarly, the voltage loop

54 *Power electronics for electric vehicles*

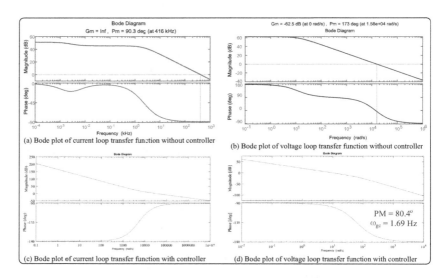

Figure 2.14 Bode plots of the PFC converter's transfer functions.

maintains a bandwidth of ~5 Hz. Bode plots of the PFC converter with a controller and without a controller are illustrated in Figure 2.14. Similarly, the same procedure is adopted for the controller design for the PSFB converter [40].

2.5.4 Simulation studies and discussion

The boost PFC-based or interleaved boost PFC-based chargers are cascaded with a PSFB converter to form a two-stage OBC. Simulation studies are performed on the MATLAB/Simulink platform with the calculated component values for the specifications mentioned in Table 2.8.

The universal input voltage, i.e., 85 V_{RMS} to 264 V_{RMS}, is considered an input, and the battery is modelled with the equivalent resistive load. The response of the conventional boost PFC-based OBC for an input of 264 V_{RMS}, 50 Hz is presented in Figure 2.15. Figure 2.15a illustrates the response of the PFC stage, i.e., input current and DC link voltage. Input current is scaled five times to improve visibility, and it is in phase with the grid voltage. The PFC converter maintains ripple-free DC link voltage at the output terminals, i.e., 400 V. Figure 2.15b illustrates the output voltage of the PSFB converter.

Similar studies are performed at the 85 V, 50 Hz, and the corresponding results are presented in Figure 2.16. A detailed comparative analysis between boost PFC-based OBC and interleaved boost PFC-based OBC in

Onboard battery charging infrastructure 55

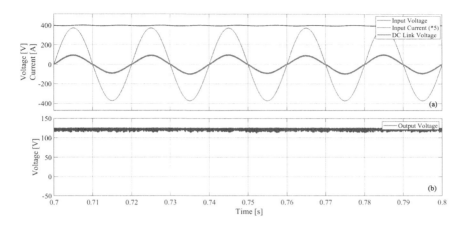

Figure 2.15 Response of the level-1 boost PFC-based OBC for 264 V input.

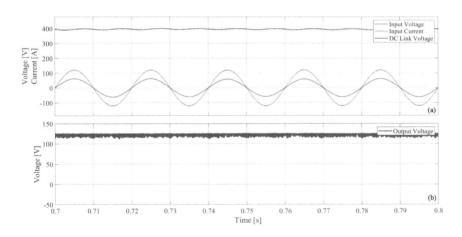

Figure 2.16 Response of the level-1 boost PFC-based OBC for 85 V input.

various aspects is depicted in Figure 2.17. Figure 2.17a concludes that both converters perform well in a wide input range and maintain nearly unity power at the input terminals. Input current ripple (Figure 2.17c), THD (Figure 2.17b), and output voltage ripple (Figure 2.17d) are within the predefined limits as specified in Table 2.8. Figure 2.17d shows the output voltage ripple for both the converters, and it's the same for both the chargers due to the same DC–DC battery interface.

56 Power electronics for electric vehicles

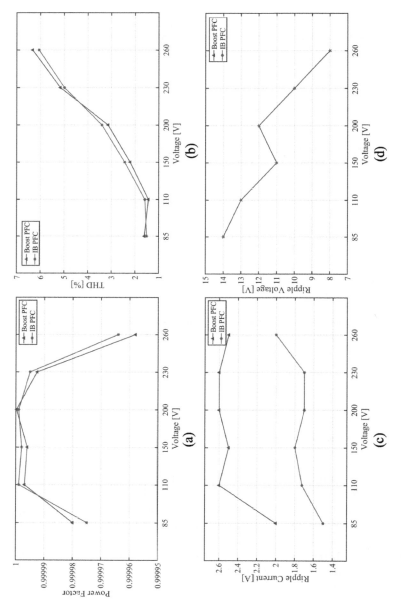

Figure 2.17 Comparative analysis of boost PFC and IB PFC-based OBCs: (a) input power factor, (b) input current THD, (c) input ripple current, (d) output voltage ripple at battery terminals.

2.6 Concluding remarks

This chapter comprehensively discussed single-phase OBCs for electrified transportation systems. Connectors, standards, and protocols associated with the EV charging systems are discussed in detail. This chapter provides in-depth knowledge of PFC converters and isolated DC–DC converters for two-stage OBCs. Two-stage OBCs are more popular due to well-established technology and are easy to control. A case study of the conventional boost PFC-based and interleaved boost PFC-based OBC is performed through MATLAB simulations. Both the chargers extract sinusoidal current at all the operating points, in phase with the input voltage. Also, they maintain the desired DC voltage at the output terminals. On the other hand, the interleaved boost PFC converter performed comparatively better. However, the bridge-type OBCs have more power loss due to the front-end bridge rectifier. Therefore, bridgeless OBCs are preferable for high-power applications. It is anticipated that advanced converter topologies with wideband gap devices will soon be predominantly used in EV chargers to achieve technical and economic benefits.

Acknowledgment

This work is supported by the Science and Engineering Research Board (SERB), the Department of Science and Technology (DST), Government of India under grant SRG/2021/000184.

References

1. "International Council of Clean Transportation," USA. Available: https://theicct.org/blog/staff/global-ice-phaseout-nov2020.
2. "Policies to Promote Electric Vehicle Deployment," USA. Available: https://www.iea.org/reports/global-ev-outlook-2021/policies-to-promote-electric-vehicle-deployment.
3. "Policies to Promote Electric Vehicle Deployment," USA. Available: https://www.iea.org/reports/global-ev-outlook-2021/policies-to-promote-electric-vehicle-deployment.
4. "Reducing the Battery Cost is Essential for Clean Energy Feature," USA. Available: https://energymonitor.ai/tech/energy-storage/reducing-battery-cost-is-essential-for-a-clean-energy-future.
5. "Types of Electric Cars and Working Principles, Omazaki Envirokal Prakarsa," Indonesia. Available: https://www.omazaki.co.id/en/types-of-electric-cars-and-working-principles.
6. "IEC 61581-1:2017," International Electrotechnical Commission, Switzerland. Available: https://webstore.iec.ch/publication/33644.
7. "Global EV Outlook," France. Available: https://www.iea.org/reports/global-ev-outlook-2021.
8. S. Rivera, S. Kouro, S. Vazquez, S. M. Goetz, R. Lizana and E. Romero-Cadaval, "Electric Vehicle Charging Infrastructure: From Grid to Battery," *IEEE Industrial Electronics Magazine*, vol. 15, no. 2, pp. 37–51, June 2021.

58 *Power electronics for electric vehicles*

9. M. Yilmaz and P. T. Krein, "Review of Battery Charger Topologies, Charging Power Levels, and Infrastructure for Plug-In Electric and Hybrid Vehicles," *IEEE Transactions on Power Electronics*, vol. 28, no. 5, pp. 2151–2169, May 2013.

10. S. Haghbin, S. Lundmark, M. Alakula and O. Carlson, "Grid-Connected Integrated Battery Chargers in Vehicle Applications: Review and New Solution," *IEEE Transactions on Industrial Electronics*, vol. 60, no. 2, pp. 459–473, Feb. 2013.

11. I. Subotic, N. Bodo, E. Levi, B. Dumnic, D. Milicevic and V. Katic, "Overview of Fast On-Board Integrated Battery Chargers for Electric Vehicles Based on Multiphase Machines and Power Electronics," *IET Electrical Power Applications*, vol. 10, no. 3, pp. 217–229, 2016.

12. D. Ronanki and S. S. Williamson, "Modular Multilevel Converters for Transportation Electrification: Challenges and Opportunities," *IEEE Transactions on Transportation Electrification*, vol. 4, no. 2, pp. 399–407, June 2018.

13. D. Ronanki, A. Kelkar and S. S. Williamson, "Extreme Fast Charging Technology–Prospects to Enhance Sustainable Electric Transportation," *MDPI-Energies*, vol. 12, no. 19, pp. 3721, Sept. 2019.

14. H. Tu, H. Feng, S. Srdic and S. Lukic, "Extreme Fast Charging of Electric Vehicles: A Technology Overview," *IEEE Transactions on Transportation Electrification*, vol. 5, no. 4, pp. 861–878, Dec. 2019.

15. A. Khaligh and S. Dusmez, "Comprehensive Topological Analysis of Conductive and Inductive Charging Solutions for Plug-In Electric Vehicles," *IEEE Transactions on Vehicular Technology*, vol. 61, no. 8, pp. 3475–3489, Oct. 2012.

16. D. Ronanki, P. S. Huynh and S. S. Williamson, "Power Electronics for Wireless Charging of Future Electric Vehicles," in *Emerging Power Converters for Renewable Energy and Electric Vehicles*, pp. 73–110, CRC Press, 2021.

17. P. S. Huynh, D. Ronanki, D. Vincent and S. S. Williamson. "Overview and Comparative Assessment of Single-Phase Power Converter Topologies of Inductive Wireless Charging Systems," *Energies*, vol. 13, no. 9, p. 2150, 2020.

18. "IEC 61851–1," Switzerland. Available: https://webstore.iec.ch/publication/33644.

19. "IEC 62196-1:2014," Switzerland. Available: https://webstore.iec.ch/publication/6582.

20. "ISO 15118-1:2013," Switzerland. Available: https://www.iso.org/standard/55365.html.

21. "SAE Electric Vehicle and Plug-in Hybrid Electric Vehicle Conductive Charge Coupler," USA. Available: https://www.sae.org/standards/content/j1772_201001.

22. J. Boyd, "China and Japan Drive a Global EV Charging Effort: The New Standard Will Be Backward Compatible With Select Charging Stations - [News]," *IEEE Spectrum*, vol. 56, no. 2, pp. 12–13, Feb. 2019.

23. "Level 1 and 2 AC Charging (Pile) Station Design Considerations," USA. Available: https://training.ti.com/level-1-and-2-ac-charging-pile-station-design-considerations?cu=1128055.

Onboard battery charging infrastructure   59

24. S. S. Williamson, A. K. Rathore and F. Musavi, "Industrial Electronics for Electric Transportation: Current State-of-the-Art and Future Challenges," *IEEE Transactions on Industrial Electronics*, vol. 62, no. 5, pp. 3021–3032, May 2015.

25. A. Khaligh and M. D'Antonio, "Global Trends in High-Power On-Board Chargers for Electric Vehicles," *IEEE Transactions on Vehicular Technology*, vol. 68, no. 4, pp. 3306–3324, April 2019.

26. H. Karneddi, D. Ronanki and R. L. Fuentes, "Technological Overview of Onboard Chargers for Electrified Automotive Transportation," in *Proceedings IECON – 47th Annual Conference of the IEEE Industrial Electronics Society*, 2021, pp. 1–6.

27. S. Habib *et al.*, "Contemporary Trends in Power Electronics Converters for Charging Solutions of Electric Vehicles," *CSEE Journal of Power and Energy Systems*, vol. 6, no. 4, pp. 911–929, Dec. 2020.

28. H. Benqassmi, J.-C. Crebier and J.-P. Ferrieux, "Comparison between Current Driven Resonant Converters Used for Single-Stage Isolated Power-Factor Correction," *IEEE Transactions on Industrial Electronics*, vol. 47, no. 3, pp. 518–524, Aug. 2002.

29. S. Chattopadhyay, V. Ramanarayanan and V. Jayashankar, "A Predictive Switching Modulator for Current Mode Control of High Power Factor Boost Rectifier," *IEEE Transactions on Power Electronics*, vol. 18, pp. 114–123, Jan 2003.

30. J. Dixom and B. Ooi, "Indirect Current Control of a Unity Power Factor Sinusoidal Current Boost Type 3-Phase Rectifier," *IEEE Transactions on Industrial Electronics*, vol. 35, no. 4, pp. 508–515, 1998.

31. D. S. Gautam, F. Musavi, M. Edington, W. Eberle and W. G. Dunford, "An Automotive Onboard 3.3-kW Battery Charger for PHEV Application," *IEEE Transactions on Vehicular Technology*, vol. 61, no. 8, pp. 3466–3474, Oct. 2012.

32. A. Dixit, K. Pande, S. Gangavarapu and A. K. Rathore, "DCM-Based Bridgeless PFC Converter for EV Charging Application," *IEEE Journal of Emerging and Selected Topics in Industrial Electronics*, vol. 1, no. 1, pp. 57–66, July 2020.

33. P. Kong, S. Wang and F. C. Lee, "Common-Mode EMI Noise Suppression in Bridgeless Boost PFC Converter," in *Proceedings APEC 07 - Twenty-Second Annual IEEE Applied Power Electronics Conference and Exposition*, 2007, pp. 929–935.

34. B. Lu, R. Brown and M. Soldano, "Bridgeless PFC Implementation Using One Cycle Control Technique," in *Proceedings Twentieth Annual IEEE Applied Power Electronics Conference and Exposition, 2005. APEC*, Vol. 2, pp. 812–817, 2005.

35. Z. Liu, F. C. Lee, Q. Li and Y. Yang, "Design of GaN-Based MHz Totem-Pole PFC Rectifier," *IEEE Journal of Emerging and Selected Topics in Power Electronics*, vol. 4, no. 3, pp. 799–807, Sept. 2016.

36. S. Y. Yu, R. Chen and A. Viswanathan, "Survey of Resonant Converter Topologies," *Texas Instruments Power Supply Design Seminar*, 2018.

37. J. J. Deng, S. Q. Li, S. D. Hu, C. C. Mi and R. Q. Ma, "Design Methodology of LLC Resonant Converters for Electric Vehicle Battery Chargers," *IEEE*

60　*Power electronics for electric vehicles*

Transactions on Vehicular Technology, vol. 63, no. 4, pp. 1581–1592, May 2014.

38. C. H. Chang, E. C. Chang and H. L. Cheng, "A High-Efficiency Solar Array Simulator Implemented by an LLC Resonant DC-DC Converter," *IEEE Transactions on Power Electronics*, vol. 28, no. 6, pp. 3039–3046, Jun. 2013.

39. Texas Instruments, "Phase-Shifted Full-Bridge DC/DC Power Converter Design Guide," TIDU248, May 2014.

40. Y. Lo, C. Lin, M. Hsieh and C. Lin, "Phase-Shifted Full-Bridge SeriesResonant DC-DC Converters for Wide Load Variations," *IEEE Transactions on Industrial Electronics*, vol. 58, no. 6, pp. 2572–2575, June 2011.

41. N. D. Weise, G. Castelino, K. Basu and N. Mohan, "A Single-Stage Dual-Active-Bridge-Based Soft Switched AC–DC Converter with Open-Loop Power Factor Correction and Other Advanced Features," *IEEE Transactions on Power Electronics*, vol. 29, no. 8, pp. 4007–4016, Aug. 2014.

42. U. R. Prasanna, A. K. Singh and K. Rajashekara, "Novel Bidirectional Single-phase Single-Stage Isolated AC–DC Converter With PFC for Charging of Electric Vehicles," *IEEE Transactions on Transportation Electrification*, vol. 3, no. 3, pp. 536–544, Sept. 2017.

43. D. Cesiel and C. Zhu, "A Closer Look at the On-Board Charger: The development of the second-generation module for the Chevrolet Volt," *IEEE Electrification Magazine*, vol. 5, no. 1, pp. 36–42, March 2017.

44. "Hyundai Motor Company," South Korea. Available: https://www.hyundai.com/content/dam/hyundai/in/en/data/connectto-service/owners-manual/kona.pdf.

45. H. V. Nguyen, D. -C. Lee and F. Blaabjerg, "A Novel SiC-Based Multifunctional Onboard Battery Charger for Plug-In Electric Vehicles," *IEEE Transactions on Power Electronics*, vol. 36, no. 5, pp. 5635–5646, May 2021.

46. "CWBC Series 6.6kW Bi-directional EV On-Board Charger," Current Ways, USA. Available: https://currentways.com/wp-content/uploads/2013/10/CWBC-Series-6.6kW-Bi-directional-OBC-Liquid-Cooled-050918.pdf.

47. "CWBC Series 6.6kW EV On-Board Charger-CAD662DF-400A," Current Ways, USA. Available: https://www.ec21.com/product-details/6.6kw-Obc--10631830.html.

48. "BRUSA," Available: http://www.brusa.biz/en/products/charger/charger-400-v/nlg513-water.html.

49. "Class-Leading Onboard Battery Charger Strengthens BorgWarners's EV Systems," BorgWarner, USA. Available: https://www.borgwarner.com/newsroom/press-releases/2019/05/16/class-leading-onboard-battery-charger-strengthens-borgwarner-s-ev-systems-leadership.

50. "Course Material on Switched-Mode Power Conversion," Indian Institute of Science, India. Available: https://www.academia.edu/download/34363079/IIT_power_book.pdf.

3 Design and development of LLC resonant converter for battery charging

Akash Gupta and Mahmadasraf A Mulla

3.1 Introduction

Electric vehicles (EVs) have several advantages in terms of environmental and economic issues over conventional internal combustion engine-based vehicles. Researchers are attempting to improve the efficiency of the charging systems even by as little as 0.1%, because the efficiency and power density are the two most important aspects of the on-board EV charging system. For the DC–DC stage of a two-stage on-board EV charger, the LLC resonant converter is the prevalent choice due to its characteristics of providing soft switching for entire load ranges, low electromagnetic interference (EMI), and integrated magnetics design with high power density and efficiency [1–6].

3.2 Working principle and configuration of LLC resonant converter

The working of a resonant DC–DC converter is based on the principle of resonance in any electric circuit. Resonant converters can have different configurations depending on the arrangement of resonant components. The LLC resonant converter is also one such converter, which has the advantages of improved efficiency, achievement of Zero Voltage Switching (ZVS) and Zero Current Switching (ZCS), better power density, and a wide range of output voltages. These converters employ frequency modulation to regulate the output voltage. The important feature of this converter is that the leakage inductance of the transformer can be utilized as series resonance inductance.

A full bridge LLC resonant converter is shown in Figure 3.1. Depending on the power rating, the input bridge can be a full bridge or a half bridge. C_r, L_r, and L_m form the resonant network. The circuit shown in Figure 3.1 can be categorized into three parts. The switches $S_1 - S_4$ of the bridge are operated in a complementary manner, which generates the square wave. This square wave is given as an input to the resonant network, which acts as a filter. It passes the fundamental and traps all the higher-order harmonics.

DOI: 10.1201/9781003248484-3

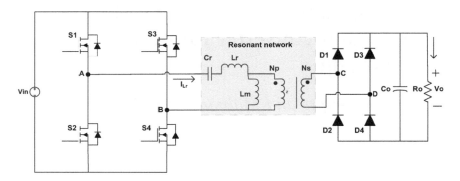

Figure 3.1 Circuit configuration of LLC resonant converter.

The power transferred by the circuit depends only on the fundamental component. The current flowing through the resonant network lags the applied voltage, which allows the switches to operate in ZVS condition. The diode bridge rectifies the alternating current (AC) at the secondary side of the transformer and provides regulated DC output.

3.3 Equivalent circuit analysis of LLC resonant converter

To obtain the gain of the LLC resonant converter, the transfer function of this converter has to be obtained. Voltage across the bridge terminals A and B is represented as V_{in_ac}, and R_{oe} represents AC equivalent resistance of combined rectifier and load resistor. To obtain the gain relation of the converter, the equivalent circuit shown in Figure 3.2 is analysed using the fundamental harmonic analysis (FHA) method, where inverter square-wave voltage is represented by the fundamental component, ignoring all higher-order harmonics [7–9]. The secondary side variables are referred to the primary side, and the effect of the output capacitor and transformer's

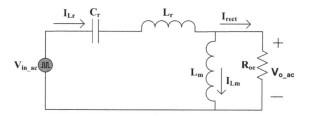

Figure 3.2 Equivalent circuit of LLC resonant converter.

LLC resonant converter for battery charging 63

secondary leakage is ignored. After applying some basic network circuit theorems, the gain equation can be written as Eq. (3.12).

From the equivalent circuit of Figure 3.2, V_{o_ac} can be written as:

$$V_{o\_ac} = \frac{V_{in\_ac}\left(\dfrac{R_{oe}.j\omega L_m}{R_{oe} + j\omega L_m}\right)}{\left(\dfrac{R_{oe}.j\omega L_m}{R_{oe} + j\omega L_m}\right) + j\omega L_r + \dfrac{1}{j\omega C_r}} \tag{3.1}$$

$$\frac{V_{o\_ac}}{V_{in\_ac}} = \frac{\left(\dfrac{R_{oe}.j\omega L_m}{R_{oe} + j\omega L_m}\right)}{\left(\dfrac{R_{oe}.j\omega L_m}{R_{oe} + j\omega L_m}\right) + j\omega L_r + \dfrac{1}{j\omega C_r}} \tag{3.2}$$

$$\frac{V_{o\_ac}}{V_{in\_ac}} = \frac{j\omega R_{oe} L_m}{j\omega R_{oe} L_m + \left(R_{oe} + j\omega L_m\right)\left(j\omega L_r + \dfrac{1}{j\omega C_r}\right)} \tag{3.3}$$

$$\frac{V_{o\_ac}}{V_{in\_ac}} = \frac{\omega^2 R_{oe} L_m C_r}{\left[\omega^2 R_{oe} C_r \left(L_m + L_r\right) - R_{oe}\right] + j\omega L_m \left[\omega^2 L_r C_r - 1\right]} \tag{3.4}$$

$$\frac{V_{o\_ac}}{V_{in\_ac}} = \frac{\dfrac{\omega^2 R_{oe} L_m C_r}{L_r}}{\left[\dfrac{\omega^2 R_{oe} C_r \left(L_m + L_r\right)}{L_r} - \dfrac{R_{oe}}{L_r}\right] + j\dfrac{\omega L_m}{L_r}\left[\omega^2 L_r C_r - 1\right]} \tag{3.5}$$

$$\frac{V_{o\_ac}}{V_{in\_ac}} = \frac{\omega^2 R_{oe} C_r \left(m - 1\right)}{\left[\omega^2 R_{oe} C_r m - \dfrac{R_{oe}}{L_r}\right] + j\omega\left(m - 1\right)\left[F_x^2 - 1\right]} \tag{3.6}$$

$$\frac{V_{o\_ac}}{V_{in\_ac}} = \frac{\omega^2 R_{oe} C_r \left(m - 1\right)}{\left[mF_x^2 - 1\right]\dfrac{R_{oe}}{L_r} + j\omega\left(m - 1\right)\left[F_x^2 - 1\right]} \tag{3.7}$$

$$\frac{V_{o\_ac}}{V_{in\_ac}} = \frac{\omega^2 R_{oe} L_r C_r \left(m - 1\right)}{\left[mF_x^2 - 1\right]R_{oe} + j\omega L_r \left(m - 1\right)\left[F_x^2 - 1\right]} \tag{3.8}$$

64 *Power electronics for electric vehicles*

$$\frac{V_{o\_ac}}{V_{in\_ac}} = \frac{F_x^2 R_{oe}(m-1)}{\left[mF_x^2 - 1\right]R_{oe} + j\omega L_r(m-1)\left[F_x^2 - 1\right]} \tag{3.9}$$

$$\frac{V_{o\_ac}}{V_{in\_ac}} = \frac{F_x^2(m-1)}{\left[mF_x^2 - 1\right] + j\dfrac{\omega L_r}{R_{oe}}(m-1)\left[F_x^2 - 1\right]} \tag{3.10}$$

$$\frac{V_{o\_ac}}{V_{in\_ac}} = \frac{F_x^2(m-1)}{\left[mF_x^2 - 1\right] + jQF_x(m-1)\left[F_x^2 - 1\right]} \tag{3.11}$$

$$K(Q,m,f_x) = \frac{V_{o\_ac(s)}}{V_{in\_ac(s)}} = \frac{F_x^2(m-1)}{\sqrt{\left(mF_x^2 - 1\right)^2 + Q^2 F_x^2 (m-1)^2 \left(F_x^2 - 1\right)^2}} \tag{3.12}$$

where,

$$Q = \frac{\sqrt{L_r / C_r}}{R_{oe}} \qquad \text{Quality factor} \tag{3.13}$$

$$R_{oe} = \frac{8N_p^2 R_o}{N_s^2 \pi^2} \qquad \text{Reflected load resistance} \tag{3.14}$$

$$F_x = \frac{f_s}{f_r} \qquad \text{Normalized switching frequency} \tag{3.15}$$

$$f_r = \frac{1}{2\pi\sqrt{L_r C_r}} \qquad \text{Resonant frequency} \tag{3.16}$$

$$m = \frac{L_m + L_r}{L_r} \qquad \text{Inductance ratio} \tag{3.17}$$

The gain curves shown in Figure 3.3 are plotted using Eq. (3.12), where gain (K) is plotted with respect to normalized switching frequency (F_x) for different values of quality factor (Q) and a fixed value of inductance ratio ($m = 6$). High values of Q represent heavy load.

The gain curves shown in Figure 3.3 are peaky in shape, and the peak defines the boundary between the capacitive and inductive regions. ZVS is always achieved in the inductive region. When the converter operates in the capacitive region, the current flowing through MOSFET (Metal Oxide Semiconductor Field Effect Transistor) leads the voltage, so it reverses its direction before the MOSFET turns off. This reverse current flows through

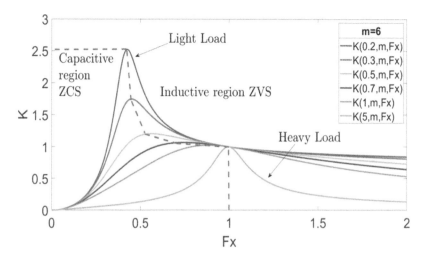

Figure 3.3 Gain characteristics of LLC resonant converter at inductance ratio $(m) = 6$.

the body diode of MOSFET and causes hard commutation in the body diode of the other switch in the same leg when it is turned ON. It may result in device failure due to increased reverse recovery losses, noise and high current spikes. Thus, the LLC resonant converter is operated in the inductive region, and the capacitive region is usually avoided. Further, the inductive region has different operating points depending on the switching frequency.

3.4 Modes of operation

The frequency modulation method is used to vary the gain of the converter. The gain of the resonant converter varies by varying the switching frequency. Depending on the load and operating switching frequency, there are three modes of operation of the converter: at resonant frequency $(f_s = f_r)$, below resonant frequency $(f_s < f_r)$ and above resonant frequency $(f_s > f_r)$ operation. Although there are three modes, the converter has only two possible switching cycle operations, as illustrated in the following. Each of the aforementioned modes may contain one or both of these operations.

3.4.1 Power delivery mode

The power delivery operation occurs twice in one switching cycle: when the positive voltage is applied to the resonant tank and when the negative voltage is applied to the resonant tank. The current resonates in the

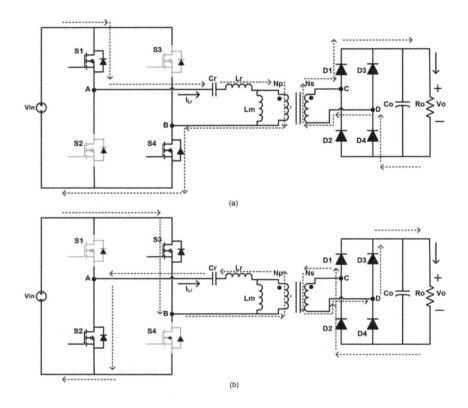

Figure 3.4 Power delivery mode of LLC resonant converter: (a) when S_1-S_4 are conducting; and (b) when S_2-S_3 are conducting.

positive direction during the first half cycle and in negative direction during the second half cycle of the switching frequency, as shown in Figure 3.4a and Figure 3.4b, respectively.

During the power delivery mode, the charging or discharging current flows through the magnetizing inductance depending on the reflected output voltage across it. The difference of the current flowing through resonant inductance and magnetizing inductance flows through the primary, which is responsible for power transfer to load.

3.4.2 Freewheeling mode

Freewheeling operation occurs when the current flowing through the resonant inductance and magnetizing inductance becomes equal during power delivery operation, as shown in Figure 3.5. This operation mainly happens

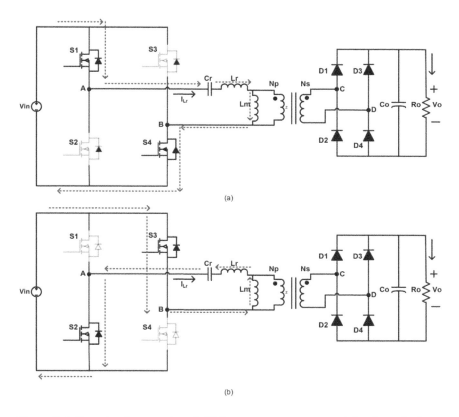

Figure 3.5 Freewheeling mode of LLC resonant converter: (a) when S_1–S_4 are conducting; and (b) when S_2–S_3 are conducting.

when the switching frequency is less than the resonant frequency ($f_s < f_r$), causing the transformer secondary side current to become zero. During this operation, the magnetizing inductor can resonate with the series resonant inductor and capacitor, and the frequency of this resonance is less than the series resonant frequency. At larger values of m ($L_m \gg L_r$), there will be a slight change in the primary current in the freewheeling operation, which can be approximated.

3.5 Design considerations

This section explains how the design parameters impact the performance, efficiency and voltage regulation of the LLC resonant converter. The objective of the design procedure is to achieve the best performance while

68 Power electronics for electric vehicles

achieving the desired gain required for all the operating conditions of line and load. The design steps are summarized as follows:

Step 1: Select Q_{max} value

Quality factor value depends on load current. High Q value represents heavy load, while low Q value is associated with a lighter load. In Figure 3.6, it can be seen that the curve with $Q = 1$ cannot reach the upper gain of 1.2, and the curve with $Q = 0.3$ reaches a lower gain of 0.8 at higher switching frequencies. Thus, the value of Q_{max} is chosen in such a way that the range of desired gain is met with less variation in switching frequency. A moderate value of Q_{max} is chosen, and the value of the inductance ratio (m) is taken as the design iteration instead of the quality factor (Q) value.

Step 2: Select m value

As discussed earlier, m is a fixed parameter, which is optimized to start the design of the LLC converter. Figure 3.7 shows the same gain curves with different values of m. Curves with low m values can reach higher gains with narrow frequency bandwidth but increase the circulating current due to the low value of magnetizing inductance. Curves with high m values cannot reach higher gains, but larger magnetizing inductance reduces the circulating current and increases efficiency. So, a reasonable m value (6 to 10) has to be chosen, which then must be optimized by some iteration to get the maximum m value that can still provide the maximum gain.

Step 3: Determine minimum value of normalized switching frequency $\left(F_{x\_min}\right)$

After the selection of a Q_{max} value and an initial value for m, the minimum normalized switching frequency will make sure that the converter is operating in the inductive region for the Q_{max} value (maximum

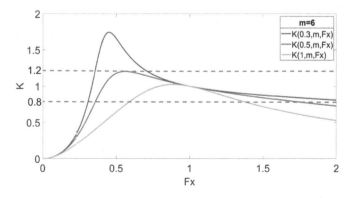

Figure 3.6 Gain curves for different values of quality factor (Q).

LLC resonant converter for battery charging 69

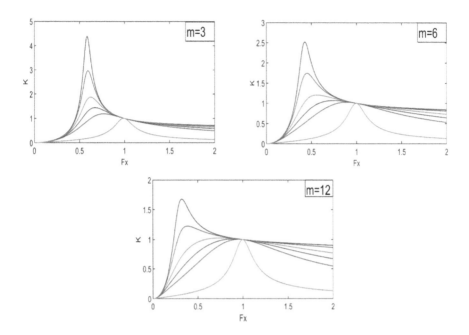

Figure 3.7 Gain characteristics of LLC converter for different values of inductance ratio (*m*).

load). The minimum normalized switching frequency occurs at the peak of the Q_{max} curve. This frequency can be found by differentiating the gain equation given by Eq. (3.12) with respect to normalized switching frequency (F_x) and equating it to zero for some fixed value of Q_{max} and *m*.

$$\frac{d}{dF_x}K(Q,m,F_{xmin}) = 0 \qquad (3.18)$$

Solving Eq. (3.18) for $Q_{max} = 0.4$, *m* = 6 will give the value of F_{x_min}.

Step 4: Verify maximum voltage gain value (K_{max})

The value of voltage gain can be verified by solving Eq. (3.19) to ensure that the maximum gain K_{max} is reached for the selected values of *m* and *Q*. If the K_{max} value is not sufficient as per requirement, then reduce the value of m and repeat step 3 and step 4 to obtain a higher boost gain.

$$K_{max} = K(Q_{max}, m, F_{xmin}) \qquad (3.19)$$

70 *Power electronics for electric vehicles*

Step 5: Calculate resonant components' values

After finalizing the value of m, one can proceed to calculate the resonant tank components' values. Eqs (3.13) to (3.17) are solved to find the values for resonant inductor (L_r), resonant capacitor (C_r) and magnetizing inductor (L_m).

3.5.1 Design example

The design steps can now be utilized to calculate the value of components required to design an LLC resonant converter prototype. The LLC resonant converter of 375 W is designed for the resonant frequency of 50 kHz. As a power supply, it is rated for an output voltage of 75 V and 5 A with the variation in input DC voltage from 230 to 350 V. The nominal input voltage is chosen as 300 V. So, the required minimum and maximum gains to be provided by the resonant circuit can be defined as

$$M_{min} = \frac{\text{Nominal input voltage}}{\text{Maximum input voltage}} \tag{3.20}$$

$$M_{max} = \frac{\text{Nominal input voltage}}{\text{Minimum input voltage}} \tag{3.21}$$

The minimum and maximum gains obtained using these relations are 0.85 and 1.3, respectively. The inductance ratio (m) value is selected as 4.5. The value of Q_{max} is chosen to be 0.33. As the nominal input voltage is 300 V, and the output voltage is 75 V, the calculated value of the transformer turns ratio (n) is 4. The main design components in the LLC resonant converter that constitutes the resonant network are series resonant inductance (L_r), magnetizing inductance (L_m) and resonant capacitor (C_r). The values of these resonant components are calculated using Eqs (3.13) to (3.17).

To calculate the values of L_r and C_r, the value of R_{oe_min} should be calculated first. To calculate the value of R_{oe_min}, substitute turns ratio $(N_p / N_s) = 4$, $R_o = 15$ in Eq. (3.14). The calculated value of R_{oe_min} after substituting all the values is 194.53 Ω. Now, substitute the value of $Q_{max} = 0.33$ in Eq. (3.13) and the value of $f_r = 50$ kHz in Eq. (3.16), and solve Eqs (3.13) and (3.16) to calculate the values of (L_r) and (C_r). The calculated values of L_r and C_r on solving Eqs (3.13) and (3.16) are obtained as 195 μH and 47 nF, respectively. Now, to calculate the value of L_m, substitute the values of $m = 4.5$ and $L_r = 195$ μH in Eq. (3.17).

The calculated value of L_m on solving Eq. (3.17) is obtained as 700 μH. Thus, the value of the resonant tank components that are the major design components of the LLC resonant converter are calculated by using the design steps presented in Section 3.5. The working of the designed resonant converters mainly depends on the transformer, as it constitutes the magnetic

LLC resonant converter for battery charging

components of the resonant converter, i.e., series resonant inductance and magnetizing inductance. The transformer is the major design part of the LLC resonant converter. The detailed design procedure is presented in the Appendix.

3.6 Control strategies of LLC resonant converter

From the operation, it is known that the output voltage or current can be regulated by regulating the switching frequency. To implement the frequency control, PFM control is employed. It is similar to pulse width modulation control in conventional DC–DC converters except that the control variable is frequency instead of duty cycle. The LLC resonant converter is operated as the power supply and a CC-CV charger. To operate it as power supply, the battery can be replaced by a resistive load. Voltage control for power supply application and CC-CV control for battery charging application are employed.

3.6.1 Voltage control for power supply application

The closed loop control for operating the LLC resonant converter as a power supply is shown in Figure 3.8. For a power supply application, the output voltage is maintained constant. The output voltage (V_o) is sensed and passed through a low pass filter. The filtered output voltage (V_{of}) is compared with the reference voltage V_{o_ref}, and error is passed through a PI controller. As per the gain equation Eq. (3.12) and gain curves shown in Figure 3.3, frequency is the control variable, so the PI controller generates frequency as output. In the PI controller, the saturation limits for

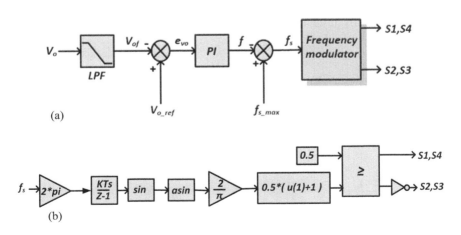

Figure 3.8 Voltage control of LLC resonant converter: (a) closed loop block diagram; and (b) expanded frequency modulator block.

72 Power electronics for electric vehicles

frequency are set from f_{s_min} to f_{s_max} as per the desired gain range. The gain plots shown in Figure 3.3 have negative slope for the inductive operating region, so the frequency obtained from the PI controller is subtracted from the maximum frequency $\left(f_{s\_max}\right)$ to get the switching frequency (f_s). Then, the frequency (f_s) is passed through the frequency modulator to generate gate pulses. This frequency modulator block is explained in Figure 3.8b. The gate pulses are then applied to the respective switches.

3.6.2 CC-CV control with separate voltage and current loops

For battery charging applications, the control loop is shown in Figure 3.9. The control strategy remains the same as that of power supply control; the only difference is that the voltage and current both need to be sensed. As the load is battery, the popular charging algorithm known as CC-CV charging is implemented. In this algorithm, the battery is charged initially with the constant current, and when the battery is charged, around 80–90% of the CC mode is switched to CV mode. As the state of charge (SOC) estimation is not part of this work, the CC to CV mode transition can be done when the voltage approaches the maximum battery voltage. The reference voltage $\left(V_{o\_ref}\right)$ is the voltage a little lower than the fully charged battery voltage. $\left(I_{o\_ref}\right)$ is the constant current with which the battery has to be charged in CC mode. A two-way switch is used to provide transition from CC to CV mode.

3.6.3 CC-CV control with cascaded voltage and current loops

In the previous CC-CV control with separate loops, separate current and voltage loops were used, i.e., both were independent. In this control, cascaded current and voltage PI loops are used, as shown in Figure 3.10. If the battery is initially uncharged, then the battery voltage is less than the set

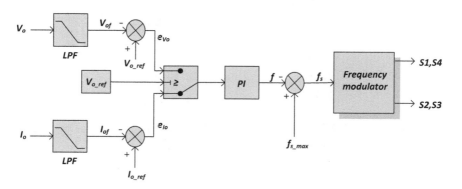

Figure 3.9 CC-CV control with separate loops for battery charging.

LLC resonant converter for battery charging

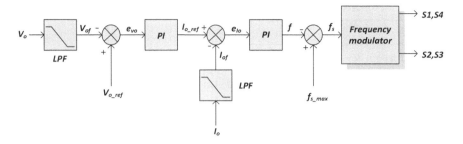

Figure 3.10 CC-CV control with cascaded loops for battery charging.

reference voltage. The battery takes some time to charge depending on the charging current and C-rating of the battery, so the error fed to the voltage PI controller cannot become zero initially. Thus, the voltage PI is operated at its upper saturation limit. The output of the voltage PI is equal to its upper saturation limit, which is kept equal to constant charging current required in CC mode. The upper and lower saturation limits of voltage PI are kept I_{o_max} and zero, respectively. The reference current $\left(I_{o\_ref}\right)$ is compared with the actual battery current and passed through the current PI controller to generate the controlled switching frequency.

3.7 Simulation study and hardware implementation

In the previous sections, the design and operation of an LLC resonant converter have been discussed. This section presents the simulation and hardware implementation of the LLC resonant converter for the power supply and battery charging applications. The LLC resonant converter is operated with the resistive load and the battery. Simulation and experimental results for the LLC resonant converter are presented by performing four different tests. In the first test, input voltage equal to nominal voltage is applied to the converter to achieve the unity gain operation. In the second test, input voltage less than nominal voltage is applied to the converter to achieve the step-up operation. In the third test, input voltage more than nominal voltage is applied to the converter to achieve the step-down operation. In the fourth test, input voltage is changed from the value below the nominal voltage to above the nominal voltage to present the behaviour of the LLC resonant converter in transient condition. In all the four cases, output voltage is maintained constant to operate the LLC resonant converter as a power supply. For each case, simulation and experimental waveforms of output voltage, output current, gate pulses, switch voltage, inductor current and secondary side current are presented. The specifications used for simulation and experimental setup are listed in Table 3.1.

74 Power electronics for electric vehicles

Table 3.1 Specifications and Parameters of LLC Resonant Converter

Parameters	Value
Resonant frequency (f_r)	50 kHz
Switching frequency range (f_s)	35–75 kHz
Gain range (K)	0.85–1.3
Transformer turns ratio (n)	4:1
Leakage inductance (L_r)	195 µH
Magnetizing inductance (L_m)	700 µH
Resonant capacitor (C_r)	47 nF
Output capacitor (C_o)	470 µF

3.7.1 Details of experimental setup

The block diagram for the hardware implementation of the LLC resonant converter is shown in Figure 3.11. AC supply is taken through a variac and applied to a rectifier to get DC supply. The DC supply is applied at the input terminals of the resonant DC–DC converter. The output voltage and output current are sensed from the load terminals using Hall voltage and current sensors. The sensed voltage and current signals are attenuated and fed to the 12-bit ADC (Analogue to Digital Converter) of a STM32F407VGT6 microcontroller. The sensed quantities are read from the ADC channels and processed by the microcontroller. The gate pulses are generated by using the advance timer. There are two advance timers in the STM32F407VGT6

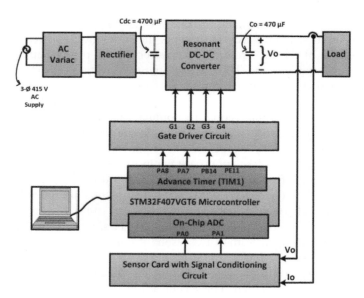

Figure 3.11 Details of experimental setup through a block diagram.

LLC resonant converter for battery charging 75

microcontroller, which are especially dedicated for power electronics application. The gate pulses are then fed to the driver circuit, consisting of TLP350 gate driver ICs (Integrated Circuits).

A lab prototype of 375 W was built to validate the design procedure and proposed control. Simulation and experimental results are presented considering 100 V as nominal input voltage for both the applications. The input voltage is varied between 75 and 115 V to resemble the gain range of 0.85 to 1.3. The performance, efficiency and experimental waveforms do not change until the converter operates within calculated values of gain.

The soft start and the closed loop control are implemented digitally using a generic 32-bit ARM Cortex-M4 microcontroller STM32F407VG. The full hardware setup for the LLC resonant converter is shown in Figure 3.13. The calculated value of resonant inductor was 185 µH, and a value of 195 µH was achieved on actual wound inductor. The resonant capacitor of 47 nF is selected, which is close to the calculated value. This combination of L_r and C_r gives resonant frequency around 53 kHz. Using Eq. (3.12), the voltage gain (M) versus normalized switching frequency (F_x) curves are plotted for different values of quality factor (Q) and inductance ratio (m) equal to 4.5, as shown in Figure 3.12. The minimum and maximum gains obtained from Eqs (3.20) and (3.21) are 0.85 and 1.3, respectively. These are represented by the two horizontal lines on gain curves, as shown in Figure 3.12 The control

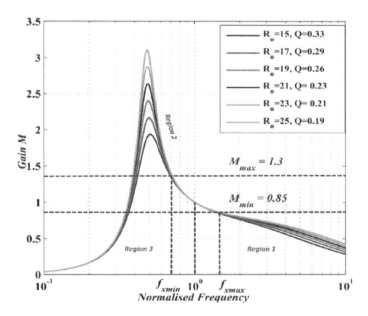

Figure 3.12 Voltage gain versus normalized frequency curves for designed LLC resonant converter.

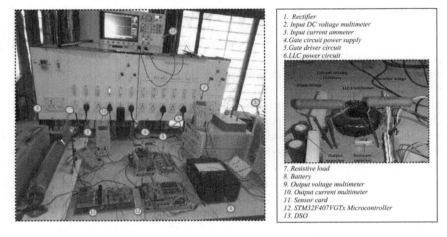

Figure 3.13 Experimental setup of LLC resonant DC–DC converter with description.

strategies have already been discussed in Section 3.6. The control equations to be implemented on hardware using the generic 32-bit ARM Cortex-M4 microcontroller STM32F407VG are discussed in the next section.

3.7.2 Control equations for hardware implementation of voltage control

For hardware implementation, the control strategy remains exactly same as discussed in Section 3.6. The output voltage is sensed using a hall sensor, and the attenuated voltage is fed to the 12-bit ADC of the STM32F407VGT microcontroller. The sensed voltage is passed through a filter given by Eq. (3.22), and the difference between actual and reference voltage is processed through a PI loop given by Eq. (3.23).

$$V_{of}[k] = V_o + 0.998\left(V_{of}[k-1] - V_o[k]\right) \tag{3.22}$$

$$f[k] = f[k-1] + k_i T_s e[k] + k_p \left(e[k] - e[k-1]\right) \tag{3.23}$$

where,

$$e[k] = V_{o\_ref}[k] - V_o f[k], k_p = 1000, k_i = 100000 \tag{3.24}$$

The PI loop is processed with the sampling time (T_s) of 1 ms, and the frequency (f) obtained is passed through the saturation limit. This frequency is then subtracted from the maximum switching frequency (f_{s_max}) as given by Eq. (3.25).

$$f_s = f_{s\_max} - f \qquad (3.25)$$

The generic 32-bit ARM Cortex-M4 microcontroller STM32F407VG has advance timers in which frequency and duty cycle can be generated directly by accessing the dedicated registers. So, the switching frequency f_s is passed to the respective timer register, which can directly generate a gate pulse. There is also direct provision for generating dead time, so the required gate pulses for all the switches can be easily generated and applied to the driver circuit and then to the respective switches.

3.7.3 Control equations for hardware implementation of CC-CV control with separate loops

For hardware implementation, the current and voltage of the battery are sensed using hall voltage and current sensors. These voltage and current signals are then read from the ADC channels and passed through the LPF filter given by Eqs (3.26) and (3.27). Both the signals are processed through the PI loops given by Eqs (3.28) and (3.29). The PI loop generates a switching frequency, which is subtracted from the maximum switching frequency.

$$V_{of}[k] = V_o[k] + 0.998\left(V_{of}[k-1] - V_o[k]\right) \qquad (3.26)$$

$$I_{of}[k] = I_o[k] + 0.998\left(I_{of}[k-1] - I_o[k]\right) \qquad (3.27)$$

$$f[k] = f[k-1] + k_i T_s e_{vo}[k] + k_p\left(e_{vo}[k] - e_{vo}[k-1]\right) \qquad (3.28)$$

$$f[k] = f[k-1] + k_i T_s e_{io}[k] + k_p\left(e_{io}[k] - e_{io}[k-1]\right) \qquad (3.29)$$

$$f_s = f_{s\_max} - f \qquad (3.30)$$

$$e_{vo}[k] = V_{o\_ref}[k] - V_{of}[k] \qquad (3.31)$$

$$e_{io}[k] = I_{o\_ref}[k] - I_{of}[k] \qquad (3.32)$$

$$K_p = 1500, K_i = 95000, T_s = 1\,\text{ms} \qquad (3.33)$$

78 Power electronics for electric vehicles

Initially, the battery is charged in CC mode, and the frequency obtained from the current loop Eq. (3.28) is passed to the timer register. When the voltage reaches the set reference value $\left(V_{o\_ref}\right)$, then CC mode is switched to CV mode, and the frequency obtained from the voltage loop Eq. (3.29) is passed to the timer register.

3.7.4 Control equations for hardware implementation of CC-CV control with cascaded loops

The battery voltage and current are sensed and passed through the low pass filter given by Eq. (3.26) and Eq. (3.27). The battery voltage (V_o) is compared with the reference voltage (V_{o_ref}), and the error is passed through the voltage PI controller given by Eq. (3.34):

$$I_{o\_ref} = I_{o\_ref}\left[k-1\right] + k_{iv}\,T_{sv}\,e_{vo}\left[k\right] + k_{pv}\left(e_{vo}\left[k\right] - e_{vo}\left[k-1\right]\right) \tag{3.34}$$

$$f\left[k\right] = f\left[k-1\right] + k_{ic}\,T_{si}\,e_{io}\left[k\right] + k_{pc}\left(e_{io}\left[k\right] - e_{io}\left[k-1\right]\right) \tag{3.35}$$

where,

$$e_{vo}\left[k\right] = V_{o\_ref}\left[k\right] - V_{cf}\left[k\right] \tag{3.36}$$

$$e_{io}\left[k\right] = I_{o\_ref}\left[k\right] - I_{of}\left[k\right] \tag{3.37}$$

$$K_{pc} = 1500, K_{ic} = 95000, T_{si} = 1\,\text{ms} \tag{3.38}$$

$$K_{pv} = 1, K_{iv} = 1.5, T_{sv} = 1\,\text{ms} \tag{3.39}$$

The designed converter was operated and analysed considering 100 V input voltage, varying it between 75 and 115 V to resemble the gain range of 0.85 to 1.3, as discussed in Section 3.5.1. When 100 V is considered as nominal input voltage, then 25 V is the nominal output voltage, and a 24 V battery can also be charged. When a 24 V lead acid battery is charged from a fixed 100 V DC input, then its voltage from uncharged to fully charged condition typically varies between 21 and 28 V; hence, the gain range falls within the design limits only. The various operating modes of the LLC resonant converter for this voltage range with experimental waveforms are presented in the next section.

3.7.5 Unity gain operation $\left(f_s = f_r\right)$

The simulation and experimental waveforms of different voltages and currents for unity gain operation of LLC resonant converter are shown in

LLC resonant converter for battery charging 79

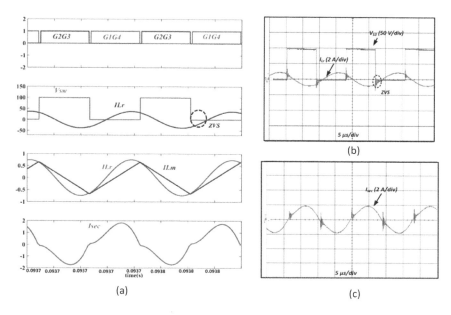

Figure 3.14 LLC resonant converter steady state waveforms at unity gain operation: (a) simulation waveforms showing gate pulses, switch voltage (V_{sw}), resonant inductor current (I_{Lr}), magnetizing inductor current (I_{Lm}) and secondary current of transformer (I_{sec}); (b) experimental waveforms of switch voltage (V_{S3}) and resonant inductor current (I_{Lr}); and (c) experimental waveform of transformer secondary current (I_{sec}).

Figure 3.14. The DC input voltage of 100 V was applied with a resistive load of 23 Ω. Figure 3.14a shows the simulation waveforms of gate pulses applied to switches $S_1 - S_4$ and $S_2 - S_3$, switch voltage (V_{sw}), current flowing through resonant inductance (I_{Lr}), magnetizing inductance (I_{Lm}) and the secondary side of the transformer (I_{sec}). The experimental waveforms of switch voltage (V_{S3}) and the current flowing through resonant inductance (I_{Lr}) and the secondary side of the transformer (I_{sec}) are obtained with similar operating conditions to those of simulation, as shown in Figure 3.14b (b) and Figure 3.14c. Figure 3.14b can be compared with simulation waveforms of switch voltage (V_{sw}) and resonant inductor current (I_{Lr}). The current through the resonant inductor (I_{Lr}) is almost sinusoidal, and ZVS is achieved. The ZVS condition is represented by a small, dotted circle in the simulation and experimental results.

Figure 3.14a and Figure 3.14c show the simulation and experimental waveforms of the transformer secondary current (I_{sec}). The transformer secondary current (I_{sec}) current is continuous and reaching zero crossing at completion of half switching cycle, hence achieving soft commutation of

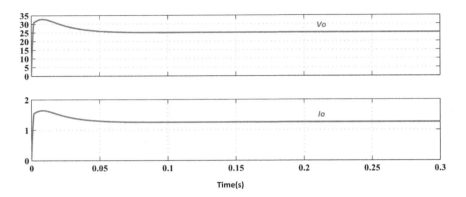

Figure 3.15 Output voltage (V_o) and output current (I_o) waveforms at unity gain operation.

secondary side diodes. The frequency of the displayed waveforms is 53 kHz, which is almost equal to the calculated value of resonant frequency for the designed components. Thus, unity gain operation is achieved at frequency equal to the resonant frequency. The simulation waveforms of output voltage and output current of LLC resonant converter at unity gain operation are shown in Figure 3.15. It can be seen that output voltage (V_o) is maintained at 25 V by the closed loop action. As the transformer turns ratio is 4:1, so at the primary of the transformer, 100 V is required to provide 25 V at the output. Now, as the applied input voltage is 100 V, it means that no step-up or step-down operation is required, and the LLC network should provide unity gain. To provide the unity gain, the PI controller regulates the frequency (f) in such a way that the switching frequency (f_s) applied to the frequency modulator is equal to the resonant frequency (f_r).

3.7.6 Step-up operation ($f_s < f_r$)

The simulation and experimental waveforms of different voltages and currents of the LLC resonant converter operating at switching frequency less than the resonant frequency for step-up operation are shown in Figure 3.16. The input voltage of 75 V was applied. The LLC resonant circuit reduces the switching frequency to increase the gain with the help of closed loop action and hence maintains the output voltage constant.

Figure 3.16a shows the simulation waveforms of gate pulses applied to switches $S_1 - S_4$ and $S_2 - S_3$, switch voltage (V_{sw}), current flowing through resonant inductance (I_{Lr}), magnetizing inductance (I_{Lm}) and the secondary side of the transformer (I_{sec}). The experimental waveforms of switch voltage (V_{S3}) and the current flowing through resonant inductance (I_{Lr}) and the secondary side of the transformer (I_{sec}) are obtained with similar

LLC resonant converter for battery charging 81

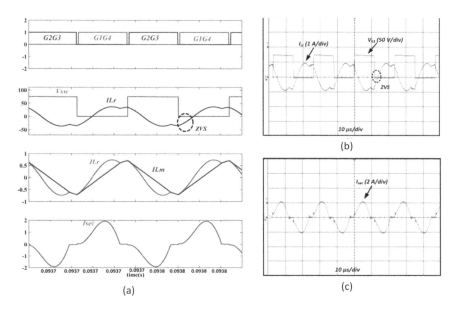

Figure 3.16 LLC resonant converter steady state waveforms at step-up operation: (a) simulation waveforms showing gate pulses, switch voltage (V_{sw}), resonant inductor current (I_{Lr}), magnetizing inductor current (I_{Lm}) and secondary current of transformer (I_{sec}); (b) experimental waveforms of switch voltage (V_{s3}) and resonant inductor current (I_{Lr}); and (c) experimental waveform of transformer secondary current (I_{sec}).

operating conditions to those of simulation, as shown in Figure 3.16b and Figure 3.16c. The experimental waveforms of switch voltage $\left(V_{S3}\right)$ and the current flowing through resonant inductance $\left(I_{Lr}\right)$ shown in Figure 3.16b can be compared with simulation waveforms of switch voltage $\left(V_{sw}\right)$ and resonant inductor current $\left(I_{Lr}\right)$. Resonant current $\left(I_{Lr}\right)$ is not as sinusoidal as it is for unity gain operation, because inductor current $\left(I_{Lr}\right)$ becomes equal to the magnetizing current $\left(I_{Lm}\right)$ before the completion of half switching cycle, i.e., prior to the turn off of switch S_1. During the interval, when $I_{Lr} = I_{Lm}$, there is no transfer of power to the secondary and circulating current flows in the primary. The ZVS condition is represented by a small, dotted circle in the simulation and experimental results.

The transformer secondary current $\left(I_{sec}\right)$ reaches zero before completion of half switching cycle, which can be seen in simulation and experimental waveforms in Figure 3.16a and Figure 3.16c, respectively, so the soft commutation of secondary diodes is also achieved. The gain in this condition is more than unity, so it is utilized for step-up operation. The output voltage and output current of the LLC resonant converter are shown in Figure 3.17 for step-up operation. In this case, the applied input voltage is less than the

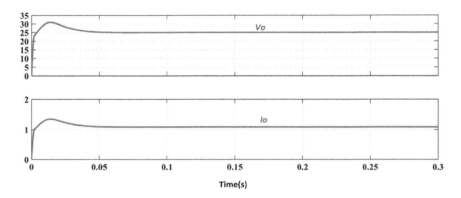

Figure 3.17 Output voltage (V_o) and output current (I_o) waveforms at step-up operation.

nominal voltage. Input DC voltage (V_{dc}) of 75 V is applied at the input of inverter bridge. It can be seen that output voltage (V_o) is maintained constant at 25 V by the closed loop action. As the applied input voltage is 75 V, this means that the LLC resonant network should step up the voltage by increasing the gain. To provide the increased gain, the PI controller regulates the frequency (f) in such a way that the switching frequency (fs) applied to the frequency modulator is lower than the resonant frequency (f_r).

3.7.7 Step-down operation ($f_s > f_r$)

The simulation and experimental waveforms of different voltages and currents of the LLC resonant converter operating at switching frequency greater than the resonant frequency for step-down operation are shown in Figure 3.18. Figure 3.18a shows the simulation waveforms of gate pulses applied to switches $S_1 - S_4$ and $S_2 - S_3$, switch voltage (V_{sw}), current flowing through resonant inductance (I_{Lr}), magnetizing inductance (I_{Lm}) and the secondary side of the transformer (I_{sec}). The experimental waveforms of switch voltage (V_{S3}) and the current flowing through resonant inductance (I_{Lr}) and the secondary side of the transformer (I_{sec}) are obtained with similar operating conditions to those of simulation, as shown in Figure 3.18b and Figure 3.18c. Figure 3.18b can be compared with simulation waveforms of switch voltage (V_{sw}) and resonant inductor current (I_{Lr}). In this condition, resonant current (I_{Lr}) does not fall to magnetizing current (I_{Lm}) when S_1 is on, so there is much less circulating current on the primary side, and there is continuous transfer of power.

The transformer secondary current (I_{sec}) does not reach zero on completion of half switching cycle, which can be seen in simulation and

LLC resonant converter for battery charging 83

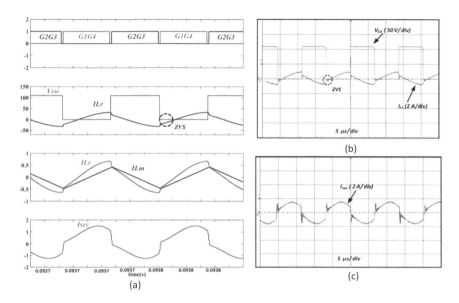

Figure 3.18 LLC resonant converter steady state waveforms at step-down operation: (a) simulation waveforms showing gate pulses, switch voltage (V_{sw}), resonant inductor current (I_{Lr}), magnetizing inductor current (I_{Lm}) and secondary current of transformer (I_{sec}); (b) experimental waveforms of switch voltage (V_{s3}) and resonant inductor current (I_{Lr}); and (c) experimental waveform of transformer secondary current (I_{sec}).

experimental waveforms in Figure 3.18a and Figure 3.18c, respectively, so soft commutation of secondary diodes is not achieved. The gain in this condition is less than unity, so it is utilized for step-down operation. The output voltage and output current of LLC resonant converter are shown in Figure 3.19 for step-down operation. In this case, applied input voltage is more than the nominal voltage. Input DC voltage of 110 V is applied at the input of the inverter bridge. It can be seen that output voltage (V_o) is maintained constant at 25 V by the closed loop action. As the transformer ratio is 4:1, 100 V is required at the primary of the transformer to provide 25 V at the output. Now, as the applied input voltage is 110 V, this means that the resonant network should step down the voltage by increasing the gain. To provide the reduced gain, the PI controller regulates the frequency (f) in such a way that the switching frequency (f_s) applied to the frequency modulator is greater than the resonant frequency (f_r).

The experimental waveforms of output voltage and output current are shown in Figure 3.20, which can be compared with the simulation waveforms shown in Figure 3.15, Figure 3.17 and Figure 3.19. For all the three operating modes, output voltage is maintained constant at 25 V by closed

84 Power electronics for electric vehicles

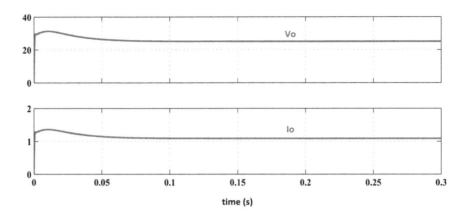

Figure 3.19 Output voltage (V_o) and output current (I_o) waveforms at step-down operation.

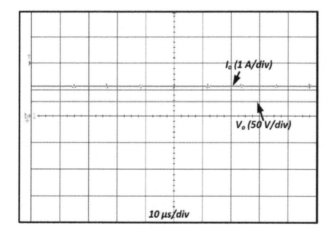

Figure 3.20 Output voltage (V_o) and output current (I_o) of LLC resonant converter as power supply.

loop frequency modulation control. Each vertical division in Figure 3.20 is 50 V for the displayed voltage waveform and 1 A for the displayed current waveform. The output voltage is 25 V, and the output current is 1.2 A.

3.7.8 Transient operation

Figure 3.21 shows the experimental waveforms for transient behaviour of the LLC resonant converter. To analyse the behaviour of the LLC resonant

LLC resonant converter for battery charging 85

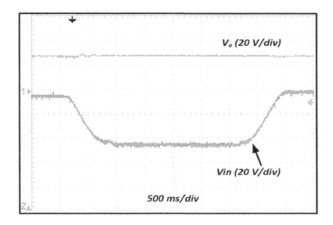

Figure 3.21 Transient operation in LLC resonant converter: output voltage (V_o); and input voltage (V_{in}).

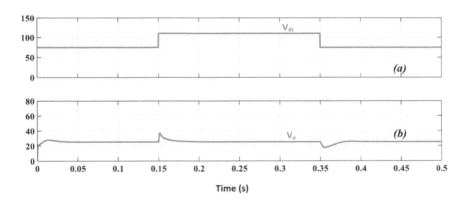

Figure 3.22 Transient operation in LLC resonant converter: (a) input DC voltage (V_{dc}) with step change at 0.15 and 0.35 s; (b) output DC voltage (V_o).

converter in transient operation, the input DC voltage (V_{in}) is changed to maintain the output voltage (V_o) constant. The input voltage is changed from 95 V to 58 V and then from 58 V to 100 V, while the output voltage is maintained constant at 25 V with the help of PI action. Figure 3.22 shows the simulation waveforms for transient behaviour of the LLC resonant converter, where line regulation is achieved. The input DC voltage V_{dc} is changed from 75 V to 110 V at 0.15 s and again from 110 V to 75 V at 0.35 s. The output voltage is maintained constant by the PI action. At both the instants of step change in input voltage, output voltage V_o is regulated back to 25 V within 0.03 s. This change in input voltage to keep the output voltage constant is called *line regulation*.

3.7.9 CC-CV charging

The simulation waveforms for battery charging in CC and CV mode are shown in Figure 3.23 and Figure 3.24, respectively. The experimental waveforms for battery charging in CC and CV mode are shown in Figure 3.25.

The lead acid battery of 24 V, 7 Ah is charged from 100 V input DC voltage (V_{dc}). When this battery is charged in CC or CV mode, it charges only with the three modes discussed earlier, so the operating waveforms will remain the same depending on gain requirement at any time. The experimental waveforms of voltage and current for battery charging with constant current of 1.5 A in CC mode are shown in Figure 3.25a. The voltage and current waveforms of the battery charging in CV mode are shown in Figure 3.25b. In CV mode, the current is reduced, and voltage equal to set reference voltage is maintained.

The simulation and experimental results of LLC resonant converter for all the three frequency regions, at resonant frequency, below resonant frequency and above resonant frequency, have been discussed. The output voltage and output current waveforms for power supply application and CC-CV mode of battery charging have also been discussed.

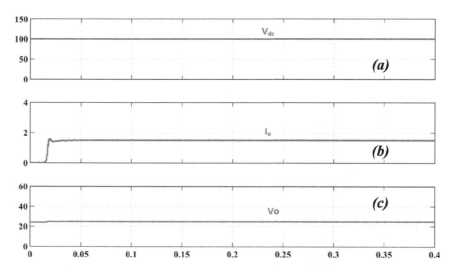

Figure 3.23 Simulation waveforms for battery charging in CC mode: (a) input DC voltage (V_{dc}); (b) output current (I_o); (c) output DC voltage (V_o).

LLC resonant converter for battery charging   87

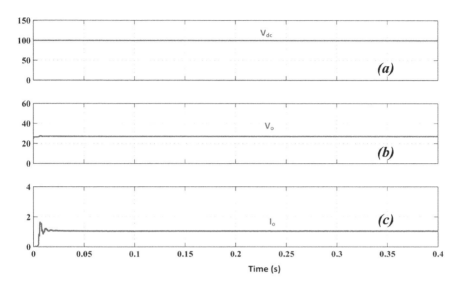

Figure 3.24 Simulation waveforms for battery charging in CV mode: (a) input DC voltage (V_{dc}); (b) output DC voltage (V_o); (c) output current (I_o).

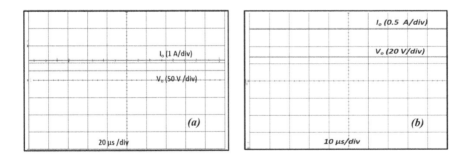

Figure 3.25 Experimental waveforms of output voltage (V_o) and output current (I_o) of LLC resonant converter: (a) battery charging in CC mode and (b) battery charging in CV mode.

3.8 Summary and conclusion

The major contribution of this work is the design, analysis and development of a resonant DC–DC converter for battery charging. The on-board charger consists of an AC–DC stage and an isolated DC–DC stage. The isolated DC–DC stage is the main focus of this work. Considering high efficiency and high power density, an LLC resonant converter is presented.

88 Power electronics for electric vehicles

Theoretical analysis and simulation are carried out for the LLC resonant DC–DC converter.

The LLC resonant DC–DC converter is presented for power supply and battery charging applications. For a power supply, a resistive load was used. A lead acid battery of 24 V, 7 Ah is charged with the very popular CC-CV charging algorithms. The step-by-step design of a high-frequency transformer with integrated leakage inductance is presented. For power supply application, voltage control is discussed. For battery charging, CC-CV control is presented in two ways, i.e., CC-CV control with separate loops and CC-CV with cascaded loops. The principle of PFM is utilized for controlling voltage and current. In simulation results and experimental results, it is observed that ZVS is achieved in the MOSFET switches. The performance of the LLC resonant converter for unity gain operation, step-up operation and step-down operation is verified through simulation results and hardware results as well.

Appendix 1: Transformer Design

To reduce the size of magnetics, the series resonant inductance is integrated in the high-frequency LLC transformer as a leakage inductance. The transformer turn ratio can be calculated by using Eq. (3.40):

$$n = \frac{\text{Nominal input voltage}}{\text{Nominal output voltage}} \tag{3.40}$$

As nominal input voltage is 300 V and nominal output voltage is 75 V, and considering the diodes voltage drop of 1.2 V also, the value of 'n' comes out to be 4. So, the turns ratio of 4:1 is used to design a transformer. The step-by-step design procedure of the LLC transformer is presented in the following:

Step 1: Area of product calculation

When the converter specifications are defined, the value of magnetizing inductance is used to calculate the area of the product. After calculating the value of magnetizing inductance (L_m) and maximum current $\left(I_{m_{peak}}\right)$ through this inductance, the inductive energy (E_L) can be calculated by using Eq. (3.41). This value of E_L is substituted in Eq. (3.42) to determine the area of product $\left(A_p\right)$ value.

$$E_L = \frac{1}{2} L_m I_{m_{peak}}^2 \tag{3.41}$$

$$A_p = A_w . A_c = \frac{2E_L}{K_w K_c J B_m} \tag{3.42}$$

LLC resonant converter for battery charging 89

where, K_w is the window factor, which is chosen as 0.4, K_c is the crest factor, which is 1, J is the current density, which is equal to 3×106 A / m^2, B_m is the maximum flux density, which is 1 T for the sendust toroidal core by KDM.

Step 2: Selection of core

A value of A_p greater than the one calculated in the preceding step is selected from the manufacturer's catalogue. Here, KS250-060A alloy powder core is chosen from ZHEJIANG NBTM KEDA MAGNETOELECTRICITY CO. LTD. (KDM) , having an area of product $\left(A_p\right)$ equal to 26541.6667 mm^2. The magnetic dimensions of the core are given in Table 3.2.

Step 3: Calculation of number of primary turns

The number of turns can be expressed in terms of permeance and inductance as:

$$N = \sqrt{\frac{L_m}{A_L}} \qquad (3.43)$$

Step 4: Check the area of product viability for the transformer

The area of product for the transformer is calculated by using Eq. (3.44):

$$A_{pt} = \frac{P_o\left(1+\dfrac{1}{\eta}\right)}{\sqrt{2}\,K_w f_s J B_m} \qquad (3.44)$$

Substituting frequency $\left(f_s\right)$ equal to 50 kHz, power $\left(P_o\right)$ equal to 375 W and efficiency equal to 0.8 in Eq. (3.44) will give A_{pt} value equal to 9943.689 mm^2. This value is lower than the calculated value of A_p, which means that the core will work fine.

Step 5: Check the primary turns viability for the transformer

Table 3.2 Magnetic Dimensions of Selected Toroidal Core

Parameters	Value
Outer diameter (D_O)	62 mm
Inner diameter (D_{in})	32 mm
Height (H)	25 mm
Cross-section area (A_c)	3.675 cm^2
Magnetic length (L_m)	14.37 cm
Window area (A_w)	7.73 cm^2
Permeance	60
AL factor (nH/Turns^2)	192

90 *Power electronics for electric vehicles*

The primary number of turns can be expressed as

$$N_p = \frac{V_{in}}{2 A_c B_m f_s}$$ (3.45)

Substituting input voltage (V_{in}) equal to 300 V and all other parameters the same as discussed previously, a number of primary turns equal to 8 is obtained, which is lower than the calculated number of turns for magnetizing inductance. Hence, the transformer will work fine with the turns calculated earlier.

Step 6: Calculation of secondary turns

The secondary number of turns can be simply calculated using the transformer turn ratio relation given by Eq. (3.46):

$$N_s = n N_p$$ (3.46)

Step 7: Wire gauge selection

The selection of wire gauge depends on the current density of the material and RMS (Root Mean Square) current flowing through it. The cross-section area of the wire is given as

$$a = \frac{i_{rms}}{J}$$ (3.47)

On substituting the current values in primary and secondary winding and the current density, the cross-section area for primary winding (a_1) is obtained as 0.833 mm², and the cross-section area for secondary winding (a_2) is obtained as 1.833 mm². According to this cross-section area, the wire gauge is selected from the SWG wire size table. SWG 20 is selected for primary winding and SWG 15 for secondary winding.

Step 8: Check for the window area availability

To accommodate the number of turns calculated in steps 5 and 6 with the wire gauge selected in step 7, there should be enough space available in the window area. To ensure this, the inequality given in Eq. (3.48) should be satisfied. If it is not satisfied, then select the next bigger size of the core and repeat the preceding steps.

$$A_w K_w > a_1 N_p + a_2 N_s$$ (3.48)

Step 9: Generating desired leakage inductance

After calculating the primary and secondary number of turns, first of all, the primary turns are wound over the entire core to get the magnetizing inductance equal to the calculated value. The secondary turns are wound loosely and as closely as possible to minimize the linkage with the primary and increase the value of leakage inductance. Leakage inductance is measured with the help of an LCR meter by shorting the secondary winding and connecting the LCR meter terminals on the primary side. This gives the value of leakage inductance, which is the sum of primary leakage and referred secondary leakage. This equivalent leakage inductance measured by the LCR meter is utilized as the series resonant inductance of the LLC resonant DC–DC converter.

References

1. Beiranvand, Reza, Bizhan Rashidian, Mohammad Reza Zolghadri, and Seyed Mohammad Hossein Alavi. "Using LLC resonant converter for designing wide-range voltage source." *IEEE Transactions on Industrial Electronics* 58, no. 5 (2010): 1746–1756.
2. Musavi, Fariborz, Marian Craciun, Deepak S. Gautam, Wilson Eberle, and William G. Dunford. "An LLC resonant DC–DC converter for wide output voltage range battery charging applications." *IEEE Transactions on Power Electronics* 28, no. 12 (2013): 5437–5445.
3. Beiranvand, Reza, Bizhan Rashidian, Mohammad Reza Zolghadri, and Seyed Mohammad Hossein Alavi. "A design procedure for optimizing the LLC resonant converter as a wide output range voltage source." *IEEE Transactions on Power Electronics* 27, no. 8 (2012): 3749–3763.
4. Deng, Junjun, Siqi Li, Sideng Hu, Chunting Chris Mi, and Ruiqing Ma. "Design methodology of LLC resonant converters for electric vehicle battery chargers." *IEEE Transactions on Vehicular Technology* 63, no. 4 (2013): 1581–1592.
5. Yang, Bo, Fred C. Lee, A. J. Zhang, and Guisong Huang. "LLC resonant converter for front end DC/DC conversion." In *APEC. Seventeenth Annual IEEE Applied Power Electronics Conference and Exposition (Cat. No. 02CH37335)*, 2, pp. 1108–1112. IEEE, 2002.
6. Tsang, Chi Wa, Martin P. Foster, David A. Stone, and Daniel T. Gladwin. "Analysis and design of LLC resonant converters with capacitor–diode clamp current limiting." *IEEE Transactions on Power Electronics* 30, no. 3 (2014): 1345–1355.
7. Abdel-Rahman, Sam. "Resonant LLC converter: Operation and design" Application Note by *Infineon Technologies North America (IFNA) Corp.* 19 (2012).
8. Huang, Hong. "Designing an LLC resonant half-bridge power converter." In 2010 Texas Instruments Power Supply Design Seminar, SEM1900, Topic, vol. 3, pp. 2010–2011. 2010.
9. Maniktala, Sanjaya. "Understanding and using LLC converters to great advantage." Application Note, Microsemi, Analog Mixed Signal Group (2013).

4 Performance evaluation of multi-input converter-based battery charging system for electric vehicle applications

R Kalpana, Kiran R, and P Parthiban

4.1 Introduction

The use of renewable energy sources (RES) is gaining popularity nowadays due to the increased cost, limited accessibility, and surplus depletion of fossil fuels. In the current world of the automotive sector, electric vehicles (EVs) have found immense importance and are making big waves in the automobile world. Moreover, the Indian market for EVs is growing exponentially, and it is expected to be around $50 billion (Rs. 3.7 Lakh crore) by the year 2030. The powerhouse of EVs is the battery pack, which comprises battery cells coupled in a combination of series and parallel to create the electricity required to run the EV and is the heart of the vehicle. In the present day, existing grid power supplies are utilized for charging these batteries. Hence, there is a necessity to provide a stable power supply to the EV charging station by utilizing RES in standalone mode conditions. This chapter proposes a multi-input converter-based hybrid power supply solution for charging EV batteries.

4.1.1 Multi-input converters

Because of environmental concerns, technological advancements, and constantly decreasing production costs, RES are becoming more popular. However, the inconsistent behaviour of the RES and the impulsiveness of the load demand is a challenging task for the elevation of these clean RES. In order to improve the dynamic and steady-state characteristics of green generation systems, power electronic converters with energy storage systems (ESS) are frequently used to convert the output power from solar photovoltaics (PV) to meet load requirements, MPPT (Maximum Power Point Tracking) control, battery integration, and impulsive energy request to handle the intermittent nature of RES. Traditionally, the RES is coupled to the load via a DC–DC converter. A bidirectional converter connects the ESS to either the input or output port of the traditional DC–DC converter for charging and discharging. The use of auxiliary converters for the ESS reduces efficiency, which is a major disadvantage of these traditional

DOI: 10.1201/9781003248484-4

methods. In addition, the multi-stage design may result in larger amplifier size, lower power density, and a higher cost. A multi-input converter (MIC) is a solution for some applications that need to combine multiple various types of input RES, such as fuel cells, wind turbines, biomass, and solar PV. With a single-stage system, this type of converter might be recommended to give the load's required power.

4.1.2 Classification of multi-input converters

A common high-voltage (HV) or low-voltage (LV) direct current (DC) bus is often used in the conventional configuration to combine several sources. For different energy sources, different DC–DC conversion phases are used. These converters are connected to the same DC bus and operate independently. Various circuit designs of MICs have been addressed in recent literature to integrate more than one RES, such as PV, wind turbine, and fuel cell, and achieve the regulated output voltage. More than one input source is included in a multiport DC–DC converter. In this setup, each input is linked to a different DC–DC converter, which is subsequently connected to a specific energy link. After that, the energy link is attached to another DC–DC converter, which is subsequently connected to the output.

Figure 4.1a depicts a basic MIC architecture. Table 4.1 compares and contrasts the performance of a traditional converter and an MIC. Isolated and non-isolated topologies are the two most common types of multiport DC–DC topologies, as shown in Figure 4.1b. Distinct DC–DC conversion stages are frequently utilized for specific sources. Those configurations are normally operated separately and are coupled together electrically at the DC bus. The location of the storage is the key structural concern of a hybrid power source (e.g., batteries). This configuration has the benefit of selecting the optimal voltage for storage. Conventionally, different converters are

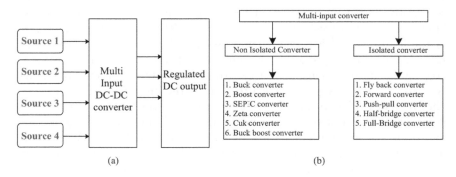

Figure 4.1 (a) Architecture of multi-input converter. (b) Classification of multi-input converter.

Evaluation of battery charging system 95

Table 4.1 Comparison of Conventional Multi-Input DC–DC Converter

Comparison	Conventional DC–DC converter	Multi-input converter
Number of power devices	High	Low
Control scheme	Distinct	Centralized control
Conversion stages	More	Fewer
Complexity	Structure complex	Control complex
Efficiency	Less	High

utilized to provide interfaces for the system's power inputs. Any basic converter topology may theoretically be used to design a power architecture for a hybrid power supply system. In comparison to the traditional way, the multiport arrangement is preferable. The number of power stages in a traditional hierarchy can be reduced by identifying severance in power processing. It is possible to separate the co-ordination resources (i.e., conversion devices) and eliminate severance in a multiport converter. Eliminating redundant power stages may lead to a higher overall system efficiency. An MIC is primarily helpful for on-site, small-scale domestic power generating systems when energy is to be obtained from various sources and energy storage is available.

4.2 Literature review

In the present era, because of limited accessibility of fossil fuels with increased price and concern over an eco-friendly environment, RES and innovation in power electronics are attracting greater attention [1–3]. It has become essential to integrate several RES, such as PV, wind, and biomass, with different capacities to a grid or standalone system through power electronic converters with low cost and compact design and efficient power flow management in distributed power systems. The conventional method comprises the integration of multiple RES separately by a single converter that is coupled to a single DC bus [4, 5]. However, because the number of powering stages is increasing, these converters are not being used effectively, resulting in increased system losses. So, in order to integrate various renewable sources, an MPC (Multi Port Converter) can be favoured over traditional converter configurations that are connected independently from sources to the DC bus. A significant literature exists on wind/solar-based hybrid systems for power generation. Since both these renewable sources are variable, it's challenging to control them both. Very few configurations have been developed to optimize the circuit configuration with consistency, effectiveness, and reduced cost. Hybrid renewable energy schemes are discussed by Faraji and Hosein [6] and Mohseni et al. [7]. Researchers have

96 *Power electronics for electric vehicles*

gone through several DC–DC converters to fulfil load demand and enhance energy from renewable sources in a hybrid system with few components. The MPCs are mainly characterized into three categories: fully isolated, partially isolated, and non-isolated [8]. Isolated and non-isolated MPCs can provide a better structure, lower price, higher power reliability, and especially high performance [9]. However, attaining ZVS (Zero Voltage Switching) in these topologies [10] is challenging, resulting in more device losses with reduced performance. As a result, an isolated MPC with an alternating current (AC) link using HFT is recommended in grid systems because of its lower price, simplicity, protection, and low leakage current [11, 12]. The three port converter (TPC) [13, 14] is considered to be the most important isolated MPC topology in the literature, having important properties such as appropriate isolation, ZVS, flexible voltage and power transfer between different RES, and so on. On the other hand, the TPC configuration is not suitable for an application in which two variable RES are combined to fulfil load requirements. This results from the fact that due to the low leakage inductance between the HFT's interconnection windings, a minor difference in magnitude and phase between their corresponding AC voltages would result in a large circulating power between the RES. As a result, increased circulating power causes more system losses. A dual-input CLL resonant configuration is presented in the literature [15], which is designed with two CLL tanks, HFTs, and a single rectifier. Although this converter design minimizes circulating currents, it suffers from load-to-source isolation. As a result, greater losses occur when the dual-input mode is used. The ports of a TPC cannot be separated by turning the switches of the converter on and off, because in the HFT, both windings are wound on the same core, and the diode of the corresponding switches that are coupled in an antiparallel arrangement provides the same passage. A half-bridge converter topology delivers half of the operational voltage on the HFT to acquire the same magnetic flux density, resulting in the HFT design being achieved with fewer primary and secondary windings. Since the number of turns on the HFT decreases, so does the DC and AC resistance. As a result, the HFT losses of a half-bridge converter topology are not much higher than those of a full-bridge converter configuration [16]. Despite the suggested design minimizing ZVS switching losses, the full-bridge configuration in all modules requires a more significant number of active switches. The power transfer capabilities of the half-bridge topology with duty-cycle adjustment are better suited than the full-bridge topology when the input voltage has large range variations due to a zero-voltage level at the HFT [17]. Furthermore, only at unity conversion ratios can simultaneous input and output ZVS operation be achieved [18–22]. This chapter introduces the dual-input half-bridge semi-active full-bridge rectifier (DIHB-SAFBR) DC–DC step-up converter to address the increased difficulties described here. This combines a dual-input half-bridge DC–DC converter with a semi-active full-bridge rectifier to provide power to the EV charging system.

Evaluation of battery charging system 97

The improved converter is intended to lower circulating power between modules by using the lowest number of passive components.

4.3 Proposed multi-input converter

Figure 4.2 illustrates the proposed DIHB-SAFBR DC–DC converter. The primary side of the proposed converter topology includes two half-bridges with four switches, S1, S2 and S3, S4, that are associated with various RES. The secondary side of the proposed configuration includes a three-leg SAFBR that is connected from the secondary of both the HFTs. It is a semi-active full-bridge converter that comprises two active legs with synchronous rectifier switches S5, S6, S7, and S8. For HFT_1, the active switches S5, S6, D_{r1}, and D_{r2} will combine to form one active full-bridge converter. Meanwhile, the S7, S8, D_{r1}, and D_{r2} have developed another full-bridge converter for HFT_2. The D_{r1} and D_{r2} form the passive leg, to reduce the quantity of switches, which shares the two active legs. The secondary of two half-bridges is connected in a parallel structure to achieve appropriate voltage regulation over a wide range. The L_{r1} and L_{r2} can either be an HFT's leakage inductance or be built independently to reach the required value and are an essential part of achieving the power handling capability. The DIHB-SAFBR design provides two HFTs, as shown in Figure 4.3. As a result, as compared with a conventional TPC, the modified configuration reduces circulating power between the modules of HFTs. Although the AC voltage of a half-bridge converter is always a square wave, the output converter's waveform is an irregular quasi-square whose shape depends on phase shift. The proposed converter configuration can be combined from various RES to perform with

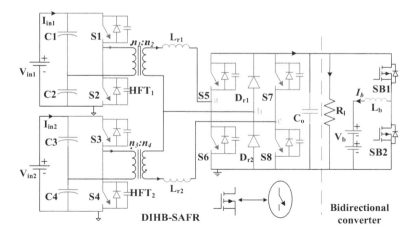

Figure 4.2 Circuit diagram of DIHB-SAFBR with bidirectional converter configuration

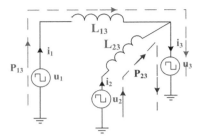

Figure 4.3 HFT equivalent circuit of the proposed system.

enhanced phase-shift ratio and magnitudes, which improves the ability to produce the required output in HFTs with significantly lower circulating current. Furthermore, ZVS is achieved over a wide range of output, resulting in increased overall system efficiency and lower losses. In addition, this chapter proposes a bidirectional DC–DC converter for battery charging and discharging. The proposed configuration employs only two active switches for battery charging and discharging, resulting in fewer devices and lower losses. The constant current/constant voltage (CC/CV) controller mode is used to charge the battery, which extends the battery's life.

4.3.1 Modes of operation for the proposed converter

When both RES are active, the proposed DIHB-SAFBR step-up converter configuration will be used as a dual-input mode, and when only one of the RES is active, it can be used as a single-input mode. The dual-input mode of operation is explored in depth in this section. Figure 4.4 shows the steady-state operational waveforms for the DIHB-SAFBR DC–DC step-up converter's operation. The operation is separated into six intervals during a complete switching cycle. Figure 4.5 depicts the respective conduction status of switches and HFT winding currents in a closed loop condition. Before beginning an examination of a new interval, it is assumed that the secondary currents i_{s1} and i_{s2} are carrying negative currents, that the active switches S2, S4, S6, and S8 are switched on, and that the rest of the switches are turned off.

$$\text{Interval 1 } (t_o < t < t_1)$$

At instant t_o, switch S2 is turned OFF as presented in Figure 4.4. During this mode of operation, the secondary voltage V_{s1} switches its polarity from −ve to +ve. Simultaneously, input current i_{p1} flows from S2 to D1 respectively in the normal manner. At instant Z_1, once secondary current i_{s1} reaches zero, switch S1 starts conducting, as depicted in Figure 4.4. Hence,

Evaluation of battery charging system 99

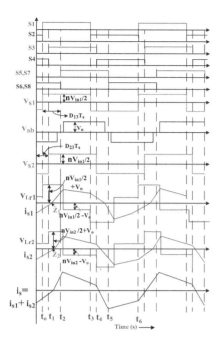

Figure 4.4 Steady-state operating waveforms for proposed converter topology.

ZVS has been achieved at this time. Figure 4.5a depicts the analogous circuit diagram for the direction of current flow in the two windings of HFTs at this time. Equations (4.1) and (4.2) indicate the secondary current response for the period, which may be derived with steady-state analysis.

$$i_{s1}(t) = \left(\frac{nV_{in1}/2 + V_o}{L_{r1}}\right)(t - t_o) + i_{s1}(t_o) \quad (4.1)$$

$$i_{s2}(t) = \left(\frac{-\frac{nV_{in2}}{2} + V_o}{L_{r2}}\right)(t - t_o) + i_{s2}(t_o) \quad (4.2)$$

Where i_{s1}, i_{s2} are secondary current of HFT_1 and HFT_2, respectively. The module input voltages are represented with V_{in1} and V_{in2}, whereas n represents HFT ratio ($n = n_2/n_1$).

Interval 2 ($t_1 < t < t_2$)

100 *Power electronics for electric vehicles*

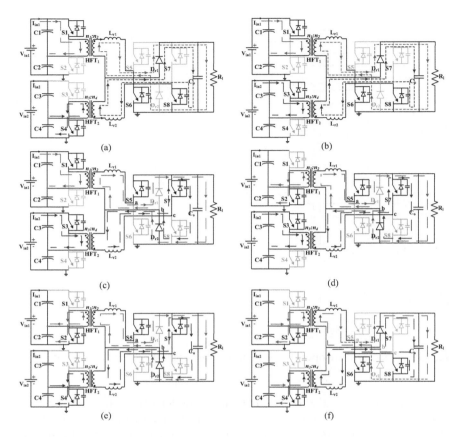

Figure 4.5 Equivalent circuits representing the current paths for the primary and secondary of HFTs during different intervals of operation of the proposed converter. (a) interval ($t_0<t<t_1$) (b) interval ($t_1<t<t_2$) (c) interval (t2<t<t3) (d) interval (t3<t<t4) (e) interval (t4<t<t5) (f) interval (t5<t<t6).

At time t_1, S4 is switched off, and secondary voltage v_{s2} is switched from −ve to +ve polarity, as indicated in Figure 4.4.

The corresponding circuit for the direction of current flow in the primary and secondary windings of HFTs at this time is depicted in Figure 4.5b. Using steady-state analysis, the secondary current expression for the interval can be determined and represented as

$$i_{s1}(t) = \left(\frac{nV_{in1}/2 + V_o}{L_{r1}}\right)(t - t_1) + i_{s1}(t_1) \qquad (4.3)$$

Evaluation of battery charging system 101

$$i_{s2}(t) = \left(\frac{nV_{in2}/2 + V_o}{L_{r2}}\right)(t - t_1) + i_{s2}(t_1) \tag{4.4}$$

Interval 3 ($t_2 < t < t_3$)

This interval starts when switches S6 and S8 have turned off at time t_2, and the secondary voltage V_{ab} transitions from –ve to +ve polarity. The secondary current, i.e., i_{s1} path, shifts from S6 to D5 and then from S8 to D7. Switches S5 and S7 are activated when currents i_{s1} and i_{s2} reach zero. The current flow direction in the primary and secondary windings of HFTs during this interval is represented with an equivalent circuit, as shown in Figure 4.5c.

The expression for the secondary current in this interval may be calculated using steady-state analysis and represented as

$$i_{s1}(t) = \left(\frac{nV_{in1}/2 - V_o}{L_{r1}}\right)(t - t_2) + i_{s1}(t_2) \tag{4.5}$$

$$i_{s2}(t) = \left(\frac{nV_{in2}/2 - V_o}{L_{r2}}\right)(t - t_2) + i_{s2}(t_2) \tag{4.6}$$

Similarly, the remaining intervals may be described using a steady-state operational waveform, as shown in Figure 4.5. An equivalent circuit for the remaining interval can be found in Figure 4.5d–f.

4.3.2 Zero-voltage switching operation of all the switches

Figure 4.4 demonstrates the key waveforms for steady-state voltage and current for zero-voltage switching. As depicted in Figure 4.6, the i_{p1} will be –ve

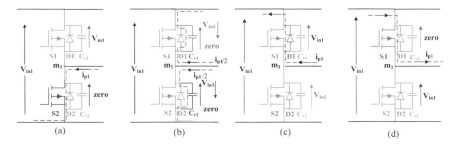

Figure 4.6 ZVS operation of S1. (a) Instant before switch is conducting. (b) Charging and discharging progression of C1 and C2. (c) Freewheeling of diode. (d) Current changing path and conduction of switch S1.

102 *Power electronics for electric vehicles*

before switch S1 is turned on. At the instant when switch S2 is switched off, the parasitic capacitance C_{r1} of switch S1 across switch drains from V_{in1} to zero. Simultaneously, the parasitic capacitor C_{r2} starts charging from zero to V_{in1}. As a result, once the charging and discharge process is complete, negative input current i_{p1} freewheels via a diode D1 of switch S1. As soon as i_{p1} changes polarity, at that instant, S1 receives a turn ON gate signal from the controller, and it begins to conduct at zero voltage. All other switches, too, work with ZVS in the same manner. The steady-state waveforms show that ZVS for input converter switches S1–S4 can only be achieved when $i_{s1}\ (t_o)$ and $i_{s1}\ (t_3)$ are less than zero.

4.4 Design of the proposed converter

i. HFT: The design of the HFT includes the selection of frequency, core, and turns ratio. Therefore, a switching frequency of f_s = 20 kHz is selected for the operation of the converter. Primary inductance of HFT is calculated as L_{p1} = 5 µH and L_{p2} = 129 µH. Similarly, secondary inductance of HFT is calculated as L_{s1} = 329 µH and L_{s2} = 302 µH by considering the A_l value from the core specification. The number of primary and secondary turns is n_1 = 30, n_2 = 240, n_3 = 30, and n_4 = 230 turns, respectively, which is required for the modified modular dual-input converter to step down voltage.

ii. Leakage inductance L_{r1} and L_{r2} of HFT can be calculated using Eq. (4.12):

$$L_r = L_{r1} = L_{r2} = \frac{nR_l}{2f_sV_o}(1-D) \tag{4.12}$$

Where n is the turns ratio, found to be 0.1 and 0.5, respectively, R_l is load, D is phase-shift ratio. So, L_{r1} is found to be 18 µH, and L_{r2} is found to be 10 µH. D is considered as 0.4.

4.5 Bidirectional DC–DC converter for battery charging and discharging application

A bidirectional buck-boost DC–DC converter is designed with the lowest possible number of active switches to charge the battery from 300 V to 48 V and discharge to satisfy the load demand, as shown in Figure 4.2. It can be perceived that it consists of an inductor with high value to reduce battery ripple and is operated with the minimum number of switches to charge and to discharge at any time as necessary according to the load. Figure 4.7 illustrates the operation of the buck and boost converter. Here, switch SB1 is utilized to charge the battery where the output voltage is stepped down from 300 to 48 V, and switch SB2 is utilized for the discharging process of the battery.

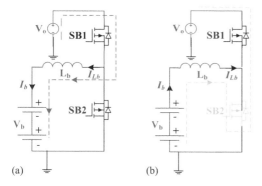

Figure 4.7. Modes of operation for bidirectional converter for (a) charging of battery and (b) discharging of battery to load.

4.5.1 Operation analysis of the proposed converter

The operation of the proposed converter can be broadly divided into two periods as detailed in this section.

Charging Mode: During charging of a battery, switch SB1, as shown in Figure 4.7a, is turned on, and switch SB2 is in OFF condition. The current I_{Lb} flows from load to charge the battery. The inductor is designed to diminish the extreme current ripple from the output end. The expression for current is represented by the following equation:

$$\frac{di_{Lb}}{dt} = \frac{V_o - V_b}{L_b}$$

Discharging mode: During discharging of a battery, switch SB2, as shown in Figure 4.7b, is turned on, and switch SB1 is in OFF condition. The current I_{Lb} flows from battery to meet the load. Here, the voltage is boosted from 48 V to 300 V. The expression for current during this mode is represented by the following equation:

$$\frac{di_{Lb}}{dt} = \frac{V_b}{L_b}$$

From Figure 4.7, it can be perceived that the path for the current flow in the inductor will be corresponding to the respective switching condition. The current can be expressed as

$$I_{Lb} = \frac{I_b}{D_{sb}} \qquad (4.15)$$

Where I_{Lb} is the average current across the inductor and D_{sb} is the duty ratio for step-down converter operation. It can be observed from Eq. (4.15) that the average current of the inductor is inversely proportional to the duty ratio of switch SB1.

Similarly, the inductor average voltage can be represented by Eq. (4.16):

$$V_b = D_{sb} V_o \qquad (4.16)$$

The voltage of the battery is V_b, and the duty ratio of the converter is D_{sb}. Step-up and step-down voltage can be attained by regulating the duty ratios. For example, if $D_{sb1} < D_{sb2}$, step-down operation can be achieved from output voltage V_o to charge the battery, and similarly, if $D_{sb2} > D_{sb1}$, step-up voltage can be gained in an effective way.

4.6 Control strategy

4.6.1 Dual-input converter

This section describes an optimal control method for the proposed dual-input converter architecture to ensure correct output voltage across the load terminals and to monitor power-sharing from source to load. The control method for the proposed converter topology is shown schematically in Figure 4.8. The ratios of phase shift D_{13}, D_{23} can be shown to play an essential role in controlling the output voltage V_o. Varying D_{13} and D_{23} [21] and the powers fed by the different RES can subsequently vary the output voltage V_o. The output voltage is maintained at the expected level by module 1. Module 2 is set up in a similar way to produce the expected output from the RES linked to it. As a result, the RES connected to module 1 will deliver the superlative required output power. In this case, D_{13} is obtained by using a proportional-integral (PI) controller on the error signal produced by comparing the standard signal with the measured output voltage. D_{23} is obtained in the same way by applying a PI controller

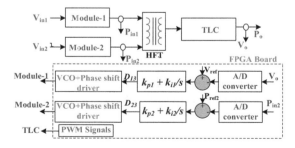

Figure 4.8 Control scheme for the proposed DIHB-SAFBR DC–DC converter.

to the error signal generated by comparing the standard and true powers given by module 1.The output voltage V_o can be controlled as desired, and the sources power flow will be achieved correctly with the proper adjustment of the phase-shift ratio and giving suitable control signals for the switches. The voltage or current controller's control signal will be sent to an FPGA controller board, which will calculate the phase-shift percentage for each source input module and give it to an isolated phase-shift gate driver. For the proposed converter, the phase-shift control is applied to regulate the output DC voltage to the reference 300 V under different load conditions. First, the difference between the V_o and the reference voltage is compared. Then, the phase-shift angle is adjusted by the PI controller to regulate V_o according to the voltage error. Due to the parameter variation of the high-frequency transformers, such as leakage inductance and turns ratio, the dual active bridge currents can be different, which results in a power unbalance of the proposed converter. A power balance control method is proposed to regulate the real power transferring through the dual-input converter. As shown in Figure 4.8, the voltage regulator compares the output voltage V_o with the reference V_{ref} and generates the error signal. Then, the power regulator compares the calculated average power of input with P_{ref2} and generates the phase-shift angles D_{23} for the proposed converter. For the PI controller gain selection, a MATLAB-based online tuning method has been preferred. For hardware implementation, DSP-based digital control is applied for achieving constant voltage across the load.

4.6.2 Control strategy for charging and discharging the battery

For optimal voltage management and voltage charging of the battery, an adequate control technique is needed. Figure 4.9 depicts a closed loop control strategy design for the proposed converter topology. Digital control is a viable approach for achieving consistent and accurate control performance. The FPGA board has been used to implement the CC/CV control algorithm. A 12-bit ADC (A/D), LPF (Low Pass Filter) and isolator gate driver, mode charge selection, a current controller, and a voltage controller are all included in the controller.

Charge of each mode CC/CV requires its own self-governing controller. Mode selection will play a vital role in selecting the mode of control according to the mode of charge. The battery voltage determines the mode of charging. If the voltage of the battery is lower when compared with the maximum charging voltage, i.e., 48 V in this chapter, the CC mode is enabled while the voltage controller is kept off. Similarly, when the battery voltage reaches its maximum charge voltage, the current controller is switched off, and the voltage controller is turned on, allowing the battery to be charged in CV mode. A PI controller, a frequency limiter, and a data processing unit make up the current or voltage controller. The PWM (pulse-width modulation)

106 *Power electronics for electric vehicles*

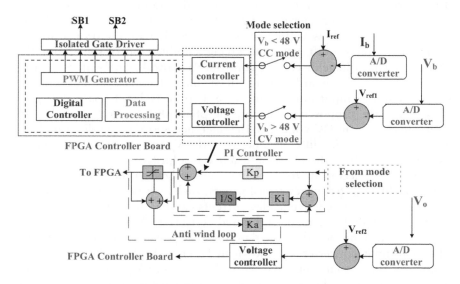

Figure 4.9 Control strategy scheme for closed loop operation of proposed dual-input converter.

signal is generated by the CC/CV controller employing the 12-bit resolution feedback voltage signal from the A/D controller.

4.7 Performance evaluation

The proposed DIHB-SAFBR DC–DC converter topology is developed for output voltage 300 V and 10 A current rating. The bidirectional DC–DC converter will charge a battery of 48 V and 6.5 Ah rating. For the proposed converter, the simulation and experimental results are discussed in this section.

4.7.1 *Simulation results*

A MATLAB simulation tool is used to investigate the accomplishment of the proposed converter topology. Figure 4.10 depicts the simulated results for the proposed converter topology. Figure 4.10a shows the waveforms of voltages of both HFTs. Here, v_{p1} and v_{s1} are voltage waveforms of HFT_1, and v_{p2} and v_{s2} are voltage waveforms of HFT_2. The waveform shows that both HFTs' input primary voltages are different, and the voltage is amplified to 300 V.

HFT_1's input and output current waveforms are shown in Figure 4.11a. It can be seen that current is appropriately shared between two modules.

Evaluation of battery charging system 107

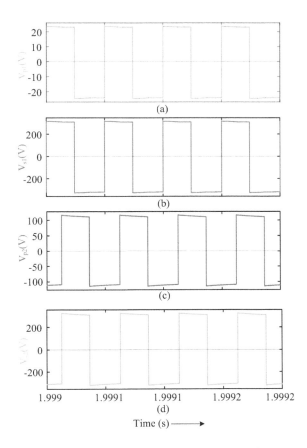

Figure 4.10 Simulated waveforms of voltage of HFTs (V_{p1}, V_{p2} are primary voltages and V_{s1}, V_{s2} are secondary voltages of HFT1 and HFT2, respectively).

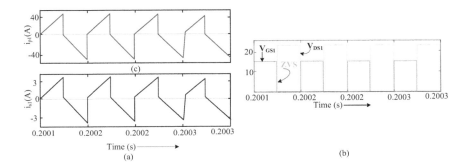

Figure 4.11 (a) Simulated waveform of input primary and secondary current of HFT1. (b) ZVS turn-on operation of Switch S1 and gate pulse of S1.

108 Power electronics for electric vehicles

The gate signals V_{gs1} for S1, switch voltage V_{ds1} and ZVS operation of the switches for the proposed converter topology are shown in Figure 4.11b. The output voltage (V_o) and current (I_o) are shown in Figure 4.12a. The output voltage appears to be ripple-free and stable regardless of load variation. It shows the effectiveness of the proposed topology under a step

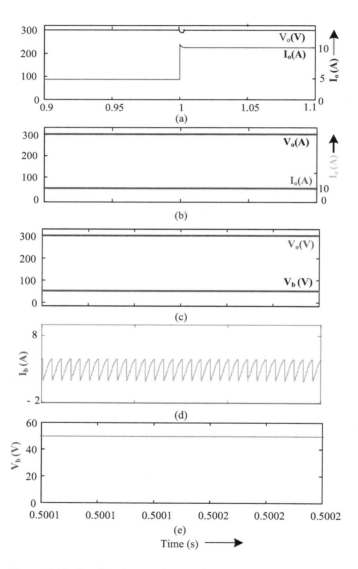

Figure 4.12 Simulated waveform of (a) dynamic behaviour under step load condition, (b) output voltage and current for 3 kW load, (c) output voltage and battery voltage, (d) battery current, (e) battery voltage.

Evaluation of battery charging system 109

load condition with proper output voltage control. Figure 4.12b represents the ripple-free voltage and current waveforms at the output side of the proposed converter. Figure 4.12c shows the voltage waveform of 300 V, which is stepped down to 48 V for battery charging purposes. The constant voltage method is utilized to charge the battery. Figure 4.12d shows the ripple current of the battery. Based on the inductor value, ripple can be minimized. Figure 4.12e shows the waveform of constant battery voltage of 48 V.

4.7.2 Experimental results

A laboratory prototype for 1 kW power is being developed to test and justify the feasibility of the functioning of a proposed converter design.

The experiment is carried out to satisfy the load demand and to charge the battery simultaneously. The proposed topology's control algorithm is demonstrated using an FPGA board. One of the input sources is assumed to give 250 W of power, while the other source delivers 750 W of power with a 230 V input voltage. Figure 4.13a and b show the input voltage/current waveform of input sources 1 and 2, respectively. The experimental voltage waveforms of both HFTs are depicted in Figure 4.14a and b. The resultant waveform illustrates that the secondary of both HFTs is stepped up to 300 V.

Voltages v_{s1} and v_{s2} appear to be trailing when compared with input voltages v_{p1} and v_{p2}, respectively. This symbolizes the flow of electricity from the input ports to the output ports. Figure 4.15 illustrates the HFTs' experimental voltage and current waveforms. Figure 4.15a and b show the secondary

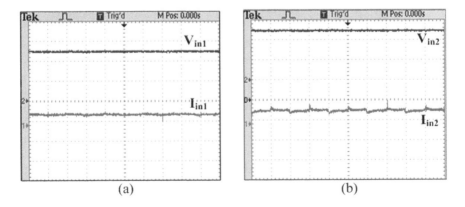

Figure 4.13 Experimental waveforms of (a) input voltage (20 V/div) and current (10 A/div) of source 1; (b) input voltage (100 V/div) and current (5 A/div) of source 2.

110 *Power electronics for electric vehicles*

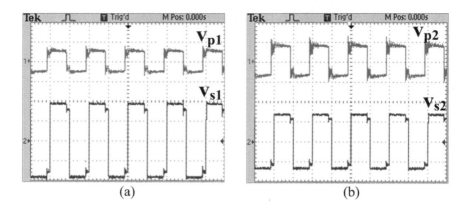

Figure 4.14 Experimental waveforms of (a) voltage waveforms of HFT$_1$ (v$_{p1}$ = 100 V/div, v$_{s1}$ = 100 V/div), (b) HFT$_2$ (v$_{p2}$ = 500 V/div, v$_{s2}$ = 200 V/div).

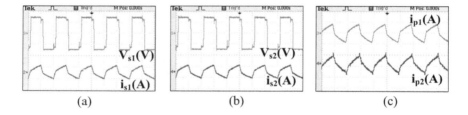

Figure 4.15 Experimental waveforms of (a) voltage and current of HFT$_1$ primary (v$_{p1}$ = 50 V/div), (b) voltage and current of HFT$_1$, (c) input current i$_{p1}$ and i$_{p2}$.

waveform of voltage and current of HFT$_1$ and HFT$_2$. We can observe that the input voltage appears to be 230 V, and current is distributed appropriately.

Figure 4.15c depicts the current waveforms of both modules. It ensures that power is shared adequately between two ports and that voltage is properly regulated. Figure 4.16 depicts the waveform of ZVS operation of switches S1 and S2. Figure 4.16b shows the gate signals (VGS1 and VGS2) applied to S1 and S2 as well as the switch voltages (VDS1 and VDS2) and the i$_{p1}$. We can perceive that at the moment at which switches S1 and S2 are turned on, the current i$_{p1}$ < 0, which satisfies the ZVS conditions. The output voltage V$_o$ and current Io are shown in Figure 4.17. The output voltage appears to be ripple-free and stable regardless of load variation. It shows the proposed topology's dynamic response under a step load condition. Figure 4.18 illustrates the experimental setup for the proposed converter. The proposed converter topology works well and efficiently with reduced

Evaluation of battery charging system 111

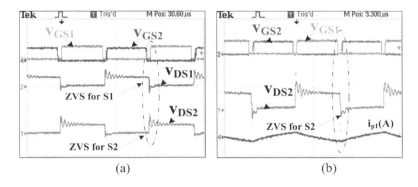

Figure 4.16 ZVS operation of switch S1 and S2 of module 1.

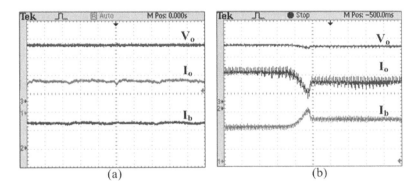

Figure 4.17 Experimental waveforms of (a) output voltage/current, (b) dynamic behaviour under step load conditions.

Figure 4.18 Experimental setup for the proposed converter with bidirectional converter.

Table 4.2 Comparison of Proposed Bidirectional Converter with Other Existing Topology

Properties (Reference)	Voltage conversion ratio (V_o/V_b), buck mode	Voltage conversion ratio (V_b/V_o), boost mode	Open current ripple (buck mode)	No of switches	No of passive devices	Magnetic component
Inverting bidirectional [23]	$-\left(\dfrac{D}{1-D}\right)$	$-\left(\dfrac{D}{1-D}\right)$	$\dfrac{V_b(1-D)}{Lf_s}$	2	3	Single inductor
Inverting bidirectional [24]	D	$\dfrac{1}{1-D}$	$\dfrac{V_b(1-D)}{Lf_s}$	4	3	Single inductor
Ćuk bidirectional converter [25]	$-\left(\dfrac{D}{1-D}\right)$	$-\left(\dfrac{D}{1-D}\right)$	$\dfrac{-V_b D}{Lf_s}$	2	5	Coupled or two inductors
Modified bidirectional with coupled inductor [26]	$\dfrac{D}{2-D}$	$\dfrac{1+D}{1-D}$	$\dfrac{V_b D}{(1+k)Lf_s}$	3	4	Coupled inductor
Proposed bidirectional converter	D	$\dfrac{1}{1-D}$	$\dfrac{V_b(1-D)}{Lf_s}$	2	3	Coupled inductor

Evaluation of battery charging system 113

circulating current and appropriate power distribution of various RES, as shown by simulation and experimental results.

4.7.3 Comparison of bidirectional converter with the proposed converter

In order to show the supremacy of the proposed converter, the following comparison has been made. Table 4.2 illustrates the supremacy of the proposed converter. From this table, it can be concluded that though SEPIC and Ćuk converter ripple current can be minimized, the number of components will be greater. So, considering all the characteristics resulting from the analysis presented, the proposed half-bridge bidirectional DC–DC is the most promising, highly efficient and robust solution for future research.

4.8 Summary

This chapter demonstrates a standalone dual-input modular-based half-bridge DC–DC converter that provides a continuous and viable power supply to an electric vehicle battery charging system. The supremacy of the proposed converter configuration is providing power to the load from both sources independently and simultaneously. It offers galvanic isolation between the source and the load, zero-voltage switch turn-on, optimal load distribution and power flow, and lower circulating currents, and reduces the risk of a magnetic short circuit, among other advantages. The proposed converter topology can be used to increase the considerable ranges of output power with regulated voltage over a wide range of phase-shift ratios between the input sources of converters. An FPGA-based DSP builder control algorithm is adopted for generating the PWM signal, which helps in modest closed loop control. A small-power prototype has been created in the laboratory for the proposed converter architecture, which is designed for 1 kW of power. The steady-state and simulation waveforms have undergone extensive testing to ensure their accuracy. According to the findings, the proposed converter architecture is well suited for applications where high voltage is required. The proposed converter configuration can be used as a promising multi-input configuration for integrating various RES and loads while achieving a controllable output with minimal losses.

References

1. Santanu Mishra, Abhishek Maji, and Soumya Shubhra Nag, "Improving Grid Power Availability in Rural Telecom Exchanges," *IEEE Transactions on Industry Applications*, vol. 54, no. 1, pp. 636–646, Jan/Feb. 2018.
2. C. W. Chen, C. Y. Liao, K. H. Chen, and Y. M. Chen, "Modeling and Controller Design of a Semi-Isolated Multi-Input Converter for a Hybrid PV/

114 *Power electronics for electric vehicles*

Wind Power Charger System," *IEEE Transactions Power Electronics*, vol. 30, no. 9, pp. 4843–4853, Sep. 2015.

3. J. Zhang et al., "PWM Plus Secondary-Side Phase-Shift Controlled Soft Switching Full-Bridge Three-Port Converter for Renewable Power Systems," *IEEE Transactions on Industrial Electronics*, vol. 62, no. 11, pp. 7061–7072, Nov. 2015.

4. D. Debnath, and K. Chatterjee, "Two-Stage Solar Photovoltaic-Based Stand-Alone Scheme Having Battery as Energy Storage Element for Rural Deployment," *IEEE Transactions on Industrial Electronics*, vol. 62, no. 7, pp. 4148–4157, Jul. 2015.

5. Rasoul Faraji, Hosein Farzanehfard, Georgios Kampitsis, Marco Mattavelli, Elison Matioli, and Morteza Esteki, "Fully Soft-Switched High Step-up Non-Isolated Three-Port DC-DC Converter Using GaN HEMTs," *IEEE Transactions on Industrial Electronics*, vol. 67, no. 10, pp. 8371–8380, Oct. 2019.

6. Rasoul Faraji, and Hosein Farzanehfard, "Soft-Switched Nonisolated High Step-Up Three-Port DC–DC Converter for Hybrid Energy Systems," *IEEE Transactions on Power Electronics*, vol. 33, no. 12, pp. 10101–10111, Dec. 2018.

7. Parham Mohseni, Seyed Hossein Hosseini, Mehran Sabahi, Tohid Jalilzadeh, and Mohammad Maalandish, "A New High Step-Up Multi-Input Multi-Output DC–DC Converter," *IEEE Transactions on Industrial Electronics*, vol. 66, no. 7, pp. 5197–5208, July 2019.

8. Xiaolu Lucia Li, Zheng Dong, Chi K. Tse, and Dylan Dah-Chuan Lu, "Single-Inductor Multi-Input Multi-Output DC-DC Converter with High Flexibility and Simple Control," *IEEE Transactions on Power Electronics*, Jun. 2020, doi: 10.1109/TPEL.2020.2991353.

9. Z. Zhang, O. C. Thomsen, M. A. E. Andersen, and H. R. Nielsen, "Dual-Input Isolated Full-Bridge Boost DC–DC Converter Based on the Distributed Transformers," *IET Power Electronics*, vol. 5, no. 7, pp. 1074–1083, 2012.

10. Erdem Asa, Kerim Colak, Mariusz Bojarski, and Dariusz Czarkowski, "Asymmetrical Duty-Cycle and Phase-Shift Control of a Novel Multiport CLL Resonant Converter," *IEEE Journal of Emerging and Selected Topics in Power Electronics*, vol. 3, no. 4, pp. 2443–2453 Dec. 2015.

11. V. Karthikeyan, and Rajesh Gupta, "Multiple-Input Configuration of Isolated Bidirectional DC–DC Converter for Power Flow Control in Combinational Battery Storage," *IEEE Transactions on Industrial Informatics*, vol. 14, no. 1, pp. 1–12, Jan. 2018.

12. R. A. Abramson, S. J. Gunter, D. M. Otten, K. K. Afridi, and D. J. Perreault, "Design and Evaluation of a Reconfigureurable Stacked Active Bridge DC/DC Converter for Efficient Wide Load-Range Operation," *2017 IEEE Applied Power Electronics Conference and Exposition (APEC)*, pp. 3391–3401, 2017, doi: 10.1109/APEC.2017.7931183

13. H. Tao, J. L. Duarte, and M. A. M. Hendrix, "Three-Port Triple-Half-Bridge Bidirectional Converter with Zero-Voltage Switching," *IEEE Transactions on Power Electronics*, vol. 23, no. 2, pp. 782–792, Mar. 2008.

14. C. Zhao et al., "An Isolated Three-Port Bidirectional DC-DC Converter with Decoupled Power Flow Management," *IEEE Transactions on Power Electronics*, vol. 23, no. 5, pp. 2443–2453, Sep. 2008.

Evaluation of battery charging system   115

15. Y. M. Chen, Y.-C. Liu, and F.-Y.Wu, "Multi-Input DC/DC Converter Based on the Multi-Winding Transformer for Renewable Energy Applications," *IEEE Transactions on Industry Applications*, vol. 38, no. 4, pp. 1096–1104, Jul./Aug. 2002.
16. Venkat Nag Someswar Rao Jakka, Anshuman Shukla, and Shrikrishna V. Kulkarni, "Flexible Power Electronic Converters for Producing AC Superimposed DC (ACsDC) Voltages," *IEEE Transactions on Industrial Electronics*, vol. 65, no. 4, pp. 3145 – 3156, 2018.
17. Yifeng Wang, Fuqiang Han, Liang Yang, Rong Xu, and Ruixin Liu, A Three-Port Bidirectional Multi-Element Resonant Converter With Decoupled Power Flow Management for Hybrid Energy Storage Systems, *IEEE Access*, vol. 6, pp. 61331–61341, 2018.
18. H. Tao, A. Kotsopoulos, J. L. Duarte, and M. A. M. Hendrix, "Transformer-Coupled Multiport ZVS Bidirectional DC–DC Converter with Wide Input Range," *IEEE Transactions on Power Electronics*, vol. 23, no. 2, pp. 771–781, Mar. 2008.
19. L. Corradini, D. Seltzer, D. Bloomquist, R. Zane, D. Maksimovic, and B. Jacobson, "Zero Voltage Switching Technique for Bidirectional DC/DC Converters," *IEEE Transactions on Power Electronics*, vol. 29, no. 4, pp. 1585–1594, Apr. 2014.
20. L. Wang, Z. Wang, and H. Li, "Asymmetrical Duty Cycle Control and Decoupled Power Flow Design of a Three-Port Bidirectional DC-DC Converter for Fuel Cell Vehicle Application," *IEEE Transactions on Power Electronics*, vol. 27, no. 2, pp. 891–904, Feb. 2012. [22] T. Zhao et al., "Voltage and Power Balance Control for a Cascaded Hbridge Converter-Based Solid-State Transformer," *IEEE Transactions on Power Electronics*, vol. 28, no. 4, pp. 1523–1532, Apr. 2013.
21. R. Kiran, and R. Kalpana, "An Isolated Dual-Input Half-Bridge DC-DC Boost Converter with Reduced Circulating Power between Input Ports," *IEEE Canadian Journal of Electrical and Computer Engineering*, doi: 10.1109/ICJECE.2021.3130723.
22. R. Kiran, and R. Kalpana, "Design and Development of Modular Dual-Input DC–DC Step-Up Converter for Telecom Power Supply," *IEEE Transactions on Industry Applications*, vol. 57, no. 3, pp. 2591–2601, May–June 2021, doi: 10.1109/TIA.2021.3056332.
23. F. Caricchi, F. Crescimbini, G. Noia, and D. Pirolo, "Experimental Study of a Bidirectional DC-DC Converter for the DC Link Voltage Control and the Regenerative Braking in PM Motor Drives Devoted to Electrical Vehicles," in *Proceedings of 9th Annual Applied Power Electronics Conference and Exposition (APEC 1994)*, Orlando, 13–17 Feb. 1994.
24. S. Waffler, and J. W. Kolar, "A Novel Low-Loss Modulation Strategy for High-Power Bidirectional Buck Boost Converters," *IEEE Transactions on Power Electronics*, vol. 24, no. 6, pp. 1589–1599, June 2009.
25. Y.-S. Lee, and M.-W. Cheng, "Intelligent Control Battery Equalization for Series Connected Lithium-Ion Battery Strings," *IEEE Transactions on Industrial Electronics*, vol. 52, no. 5, pp. 1297–1307, Oct. 2005.
26. L.-S. Yang, and T.-J. Liang, "Analysis and Implementation of a Novel Bidirectional DC–DC Converter," *IEEE Transactions on Industrial Electronics*, vol. 59, no. 1, pp. 422–434, Apr. 2011.

5 Analysis of solar pv-based electric vehicle charging infrastructure with vehicle-to-grid power regulation

Sheik Mohammed Sulthan,
Mohammed Mansoor O, and Ulaganathan M

5.1 Introduction

Worldwide energy consumption is steadily increasing every year. By the end of 2019, 419 million terajoules (TJ) of energy had been consumed by various sectors, like industries, transport, residential, etc. The transport sector is the second-largest energy consumer, taking up nearly 121 million TJ of energy [1]. Almost 90% of the transport sector's energy consumption is met from petroleum by-products. The increasing population and enormous development in the transport sector accelerated the requirement for fossil fuel and hiked its price. In addition, the transport sector is one of the top emitters of greenhouse gases, emitted 7.3 GT of CO_2 in the year 2020 [2]. Not only do fossil fuels lead to environmental pollution, but also their availability is reducing day by day. Currently, most conventional vehicles use the internal combustion engine (ICE). The ICE uses fossil fuels as a driving source and exhausts harmful CO_2 into the environment, which leads to global warming. A promising alternative to the ICE is transport sector electrification using battery-operated electric vehicles (B-EVs). B-EVs not only limit fossil fuel usage but also drastically reduce greenhouse gas emissions. EV technology has seen significant progress in the past decade. It is a promising solution to reduce CO_2 emissions and it helps to improve the power quality in power distribution systems. A typical B-EV includes an electric motor acting as propulsion energized by a storage unit, usually with a battery [3].

The role of power electronic converters in electric vehicle and charging systems is vital. Several power conversion and processing stages take place between battery charging and power drive. Figure 5.1 shows a generic block diagram that represents the power conversion stages and different types of converters used in vehicles and chargers. Every EV has an on-board charger, which helps to connect the EV for charging from alternating current (AC). The on-board charger converts the power from AC to direct current (DC) and regulates the power using a DC–DC converter in grid-to-vehicle (G2V) mode. On the other hand, ton-board charger performs DC–DC conversion, power factor correction, and DC–AC conversion in vehicle-to-grid

DOI: 10.1201/9781003248484-5

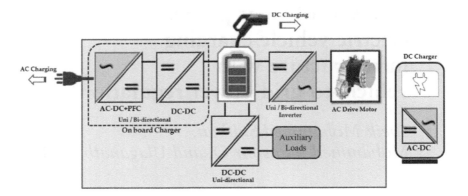

Figure 5.1 Power electronic converters in EV.

(V2G) mode. The on-board chargers with which V2G-facilitated vehicles are equipped are bi-directional. The heart of the EV is the electric motor drive equipped with the main power converter. The main power converter connecting the electric drive to the battery is bi-directional in the case of a vehicle that has regenerative braking capability. The auxiliary systems such as lighting, audio visual, and comfort are powered by an independent low-voltage converter. This implies the significance of and need for development of efficient and reliable converters.

The growth in EV technology is synergizing EV manufacturers and electricity service providers in the EV charging domain. The EV can be charged through DC or AC systems. EV batteries are currently charged by conductive and inductive charging at conventional power outlets or charging stations [4]. The EV charging station may affect the power distribution system by loading the grid due to a poorly designed charging station [5].

The electricity sources used to charge the batteries in the EV define its benefits in terms of emission reduction. EVs become emission-free only when they are charged from renewable energy sources (RES) like solar power, wind power, etc. [6]. The integration of RES into the EV charging infrastructure also introduces a few more power converters as well in the whole system. The energy storage system can stabilize the variable electricity generated by the resources by storing or supplying the electricity into the grid [7]. Similarly, the EV batteries can act as an energy storage system, enabling EVs to charge the battery through electricity from the RES and export the power to the grid, commonly named V2G operation [8].

With the increasing need for EV charging stations, solar PV-based charging facilities are gaining more attention due to sustainability and ability to support the utility grid. Feasibility studies in implementing rooftop photovoltaic (PV) systems to charge the EV storage unit showed promising

Solar PV-based charging infrastructure 119

results [9, 10]. The combined possibility of V2G operation in the EV, and power generation from PV systems, enable avenues for optimal power generation scheduling, as reported in Traube et al. [11]. The results conclude that the proposed system attains minimal grid operating cost and improves the grid operation. Detailed analysis on the EV charging point and integration of solar rooftop is discussed in Tuffner et al. [12]. The reported study reveals the interaction in the EV charging station powered by a large solar PV rooftop and distribution system. The proposed model reduced the fluctuations in grid voltage by 15% and met the grid demand through V2G support. An extensive review on the solar powered electric car charging station in the Swiss city of Frauenfeld is reported in Neumann et al. [13], which brought out the benefits of a solar power parking station that covers up to 40% of EV energy demand. Hassan Fathabadi [14] proposed a solar-powered EV charging station with V2G support and connected to the grid, including PV array, maximum power point tracking (MPPT) controller, bi-directional DC/DC converter, and bi-directional inverter.

From the preceding discussion, the need for shifting towards electric mobility, the role of power converters and their importance, and the benefits of adopting non-conventional sources for charging EVs can be realized. This chapter discusses solar-powered charging of EVs in a DC grid and power regulation of the utility grid using V2G operation. The chapter is organized as follows. Section 5.2 discusses EV charging and the role of power converters in a solar PV-based DC grid. Power regulation of a large-scale utility grid with solar PV plant and residential/commercial load using V2G power is analysed in Section 5.3, and Section 5.4 concludes the chapter.

5.2 EV charging in a Solar PV-based DC grid

The architecture of solar-powered EV charging station is shown in Figure 5.2. The entire charging station is a DC system. This small-scale DC charging station includes the PV system, an energy storage system, and the AC grid source. Different converter topologies are employed at various stages of the circuit, and they are controlled by the energy management system (EMS). To ensure the controlled and reliable operation of the system, the EMS monitors the PV power generation, the state of charge (SOC) of the energy storage system (ESS) battery, and the EV battery. The voltage level in this architecture is 48 V.

A 1 kW solar PV system is considered as the source of the charging system. The PV module has a maximum power output of 250 W_p. The open circuit voltage and short circuit current of the PV module are 37.3 V and 8.66 A, respectively. A solar PV string is formed by connecting four solar PV modules in series. The solar PV system is connected with the DC bus via an interleaved buck converter. The PV system has maximum

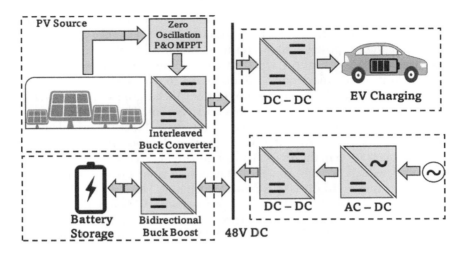

Figure 5.2 Architecture of solar-powered DC charging system.

output voltage of 120 V DC, which is stepped down to the DC charging system bus voltage level of 48 V by a two-phase interleaved buck converter. Interleaved converter is used since it reduces the ripple in the input current and output voltage. The rated power of the interleaved buck converter is 1 kW.

Under varying irradiation conditions, PV system power output is maximized by MPPT controllers. Perturb and Observe (P&O) is the most popular classical algorithm due to its simplicity and implementation feasibility [15, 16]. In this work, an improved version of P&O MPPT controller which has faster tracking and zero oscillation at the MPP, is used [17, 18, 19]. The EV battery has 48 V, 20A h capacity and is connected to the DC bus via a DC voltage regulator. A bi-directional buck-boost converter connects the battery storage system to the DC bus. The storage system battery capacity is 12 V, 1000 Wh. Hence, the converter works in buck mode (48 V–12 V) during charging and boost mode (12 V–48 V) while supplying energy to the DC bus. The rated capacity of the bi-directional converter is 1 kW. The AC input is converted using an AC–DC rectifier circuit and further regulated using a DC–DC converter to integrate the AC power source into the DC system. The AC power source supplies the required power to the DC system when the energy is available in the storage system and the power generated by the PV source doesn't meet the demand. The converter circuits are designed following the standard design specification presented in Sheik and Devaraj [20] and Thomas and Sheik [21]. All the converters are designed for the switching frequency of 10 kHz.

5.2.1 Analysis of small-scale DC charging system

This section analyse of the simulation results of the small-scale DC charging system. Here, the modified zero oscillation MPPT algorithm is used to track the peak power of the solar PV module under varying irradiation conditions. The variation in irradiation and the corresponding duty cycle observed from the proposed small-scale test system are shown in Figure 5.3 and Figure 5.4, respectively. At every 50 ms, irradiation changes with the assumption that the temperature is constant. Here, five time intervals – instant 1 to 5 – each 50 ms long are taken into account for the simulation study. The unique feature of the zero oscillation P & O MPPT algorithm is that after reaching peak power, it will remain in steady state without any oscillation. From Figure 5.4, it is very clear that the duty cycle for this MPPT algorithm starts at 0.1 and reaches a duty cycle corresponding to maximum power. Depending on irradiation variations, this algorithm adjusts the peak power by providing larger and smaller duty cycles. As a result, this MPPT algorithm guarantees MPP tracking under all conditions.

Figures 5.5 to 5.7 show the variation in PV system parameters such as voltage, current, and power generated by the PV corresponding to the

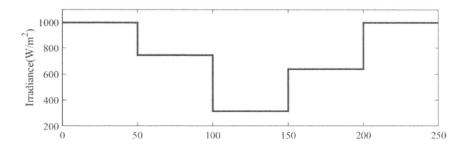

Figure 5.3 Irradiance pattern.

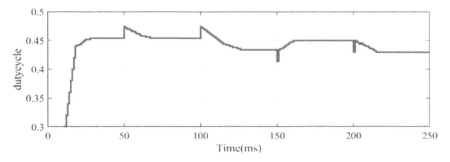

Figure 5.4 Duty cycle generated by zero oscillation MPPT algorithm.

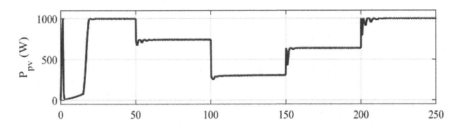

Figure 5.5 PV power.

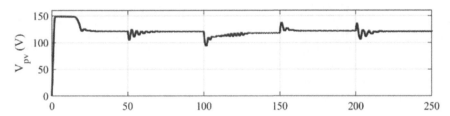

Figure 5.6 PV voltage.

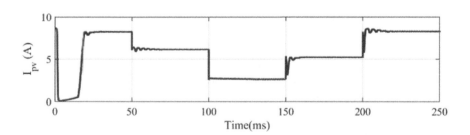

Figure 5.7 PV current.

irradiation variation. It can be seen that because of the change in irradiation, there is a rapid change in the PV current and a minor change in the PV voltage generated.

The power generated by PV, EV charging power, power in battery storage system, and grid power are illustrated in Figures 5.8 to 5.11. It is very clear from these figures that at the first, second, and fourth instants, the electric vehicle is in charge mode, and at the third and fifth instants, it is in disconnected mode. Thus, during instants 1, 2, and 4, the power flow can be seen into the EV, and in disconnected mode, the power is flowing into the storage system.

Solar PV-based charging infrastructure 123

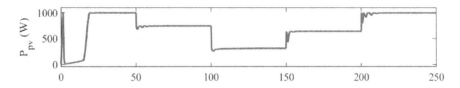

Figure 5.8 PV generated power.

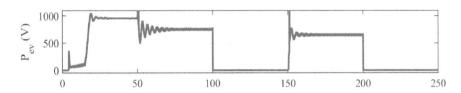

Figure 5.9 EV charging power.

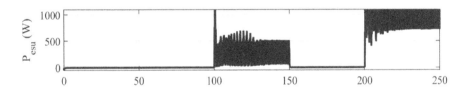

Figure 5.10 Electric storage unit power.

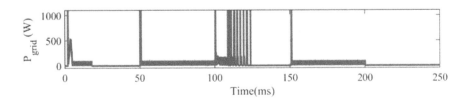

Figure 5.11 Grid power.

Thus, Figures 5.8 to 5.11 indicate that whenever there are variations in power in the system, that power is compensated for by the grid power. For example, at instants 2 (50 to 100 ms) and 4 (150 to 200 ms), the power required by the EV exceeds the power generated by PV. Thus, during these time zones, power regulation in the system is achieved by drawing power

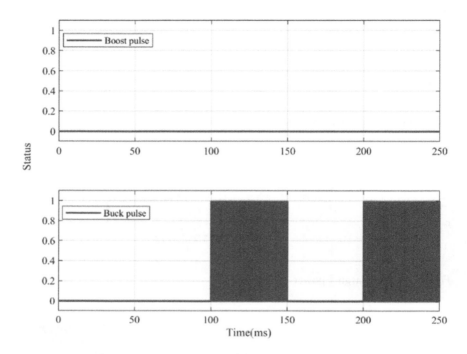

Figure 5.12 Status of boost and buck pulses.

from the grid. The boost and buck mode pulse of the bi-directional DC–DC converter is shown in Figure 5.12. In this case, the initial battery capacity of the storage system is assumed to be only 10% (SOC of the storage battery is 10%). It is to be noted that from instants 3 and 5, the EV is not charging, and PV is generating power. Consequently, the power generated by PV is used to charge ESS from 10% SOC to the value required to discharge the battery when needed to charge the EV. This can be viewed from the buck pulse of the bi-directional converter at instants 3 and 5 of Figure 5.12.

Figure 5.13 shows the SOC of the EV and ESU. From these plots, it is clear that whenever EV charging is not required (at instants 3 and 5), PV power is used to charge the electric storage unit. Therefore, the EMS adopted in the proposed DC nanogrid regulates the flow of power between the different components of the system.

5.3 Power regulation in a large-scale utility grid

Power regulation of large scale utility grid with the help of EV charging station, which supports V2G is discussed in this section. The system parameters and charging station parameters used for the study are listed in

Solar PV-based charging infrastructure 125

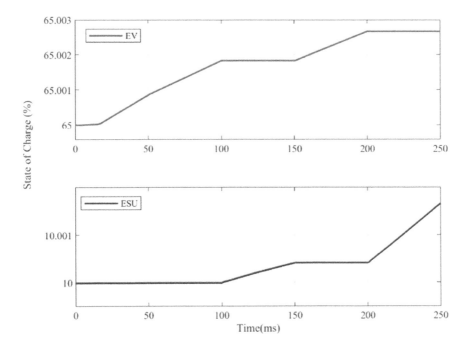

Figure 5.13 SOC of EV and ESU.

Table 5.1 System Configuration

Sl. No.	Components	Capacity (MW)
1	Solar PV Plant	1
2	Residential and Commercial Load	1.5
3	Public EV Charging Station	1.1

Table 5.1 and Table 5.2, respectively. For this study, five different charging profiles are created, and a selected number of vehicles are assigned to each group, as tabulated in Table 5.3. The charging profile data includes the plugin time, plug out time, SOC of vehicle during each interval, and travel time for each group of vehicles. G2V charging and V2G power regulation with respect to the PV generation and load conditions are analysed. This study also describes the dynamic stability of the system in the event of sudden variations in solar PV production due to changes in irradiation. The irradiation data for the proposed system is taken from the weather conditions of Kollam, Kerala, India (8.9142° N, 76.6320° E). The irradiation data of selected location on May 2, 2021 is shown in Figure 5.14.

126 Power electronics for electric vehicles

Table 5.2 Charging Station Parameters

Sl. No.	Components	Capacity
1	Number of Vehicles	50
2	Charging Power	22 kW
3	Maximum Power Capacity of Charging Converter	50 kW
4	Efficiency of Power Converter	90%

Table 5.3 Vehicle Groups

Sl. No.	Vehicles Profile	Number of Vehicles
1	Group 1	17
2	Group 2	12
3	Group 3	9
4	Group 4	5
5	Group 5	7

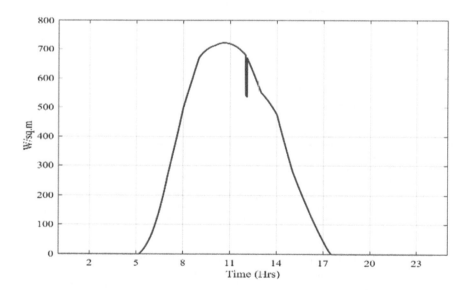

Figure 5.14 Irradiation data on May 2, 2021.

5.3.1 Case 1

Figure 5.15 shows the PV solar power generated, the total residential and commercial load pattern, the EV load, and the power output of the grid over the day with a sampling time of 1 minute. The combined power generation and consumption are shown separately in Figure 5.16. The SOC variation of the EV groups in case 1 is shown in Figure 5.17.

Solar PV-based charging infrastructure 127

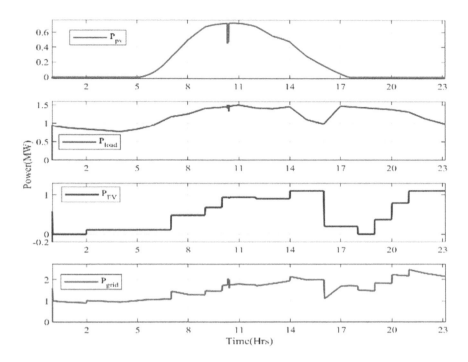

Figure 5.15 Power variation.

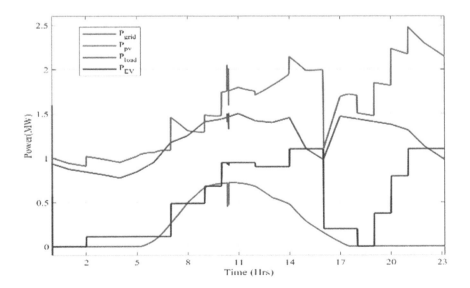

Figure 5.16 Combined power variation.

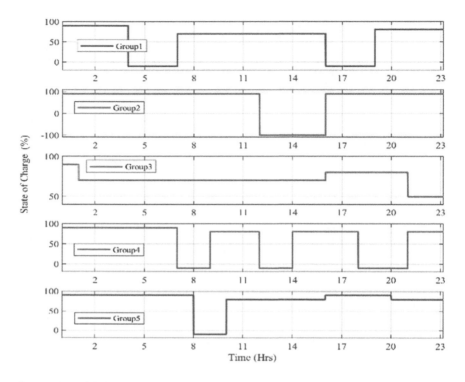

Figure 5.17 SOC variation.

The negative SOC in Figure 5.17 indicates that during that time, the respective EV groups are supplying power to the grid. The number of vehicles at any instant of time feeding power to the grid or taking power from grid is shown in Figure 5.18. It should be noted that at any time, vehicles do not need to perform charging only; some vehicles may also perform power regulation.

5.3.2 Case 2

Figure 5.19 shows the PV generation and power consumption pattern in case 2. Figure 5.20 and Figure 5.21, respectively, show the variation in the SOC of the EV groups and the number of vehicles at any time of the day.

In case 2, it is clear from Figure 5.20 and Figure 5.21 that certain groups of vehicles go nowhere; they are constantly in plugin mode. During this time, these vehicles merely regulate the power supply. The sudden interruption of solar PV production for a period of about 5 minutes is also studied in case 2. Figure 5.22 depicts the power regulation in the proposed system

Solar PV-based charging infrastructure 129

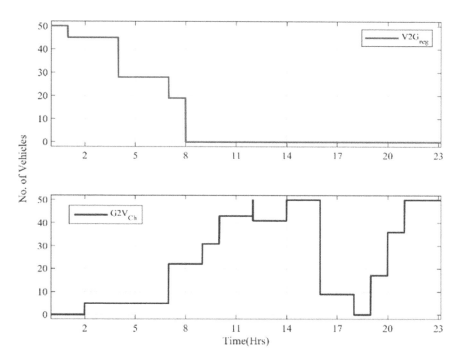

Figure 5.18 Number of vehicles in case 1.

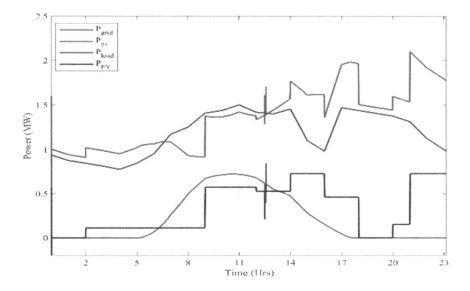

Figure 5.19 Power variation in case 2.

130 Power electronics for electric vehicles

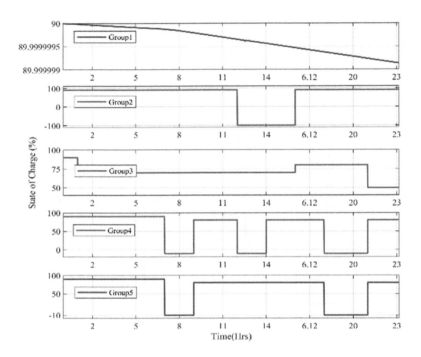

Figure 5.20 SOC variation in case 2.

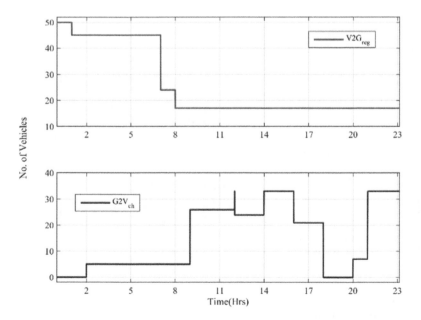

Figure 5.21 Number of vehicles in case 2.

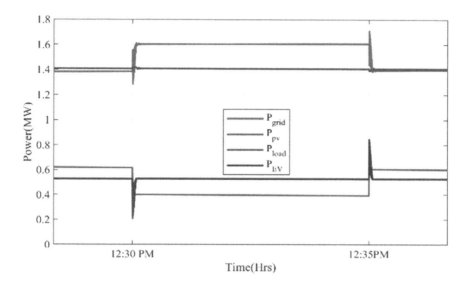

Figure 5.22 Power regulation in case 2.

during this period. In this interval, the power balance is obtained by drawing energy from the grid, i.e., G2V operation.

5.4 Conclusion

The importance and role of power electronic converters in e-mobility, a small-scale solar-powered DC charging system for EVs, and grid power regulation in a large-scale utility grid using V2G terminology are discussed in this chapter. A detailed discussion on various power converters used in EV charging stations is presented. A solar PV integrated EV charging system model is designed and built in MATLAB. Simulation of the proposed small-scale DC charging system model with solar PV is performed under various conditions, and results are obtained. Power flow between the sources with respect to PV generation and SOC level of the EV battery, status of the EV battery, and SOC level of ESS are realized from the simulation results. The effectiveness of the controller is also validated by the results. Further, grid power regulation in a large-scale utility system with a 1 MW solar power plant and 1.1 MW EV charging facility is tested. The simulation is conducted for 24 hours with real-time solar irradiation data. The model is configured with a maximum residential and commercial load of 1.5 MW; a total of 50 vehicles grouped under different profiles are considered over a period of 24 hours. Through simulation studies, the grid power regulation using the vehicle power is evaluated.

132 Power electronics for electric vehicles

Appendix

Design equations for interleaved buck converter

The converter's components and its duty cycle are calculated using Eqs (5.1)–(5.3).

$$L = \frac{V_o(1-D)}{\Delta I_L f} \tag{5.1}$$

$$C = \frac{V_o(1-D)}{8Lf^2\Delta V_C} \tag{5.2}$$

$$D = \frac{V_C}{V_S} \tag{5.3}$$

Design equations for bi-directional buck-boost converter

The components of the bi-directional converter are designed with the help of Eqs (5.4)–(5.8).

Boost mode configuration

$$V_H = \frac{1}{1-D}V_L \tag{5.4}$$

$$L_m = \frac{D(1-D)^2 R_H}{2f} \tag{5.5}$$

$$C_H = \frac{D}{R_H f\left(\dfrac{\Delta V_H}{V_H}\right)} \tag{5.6}$$

Buck mode configuration

$$V_H = \frac{V_L}{D} \tag{5.7}$$

$$C_L = \frac{V_L(1-D)}{8\Delta V_L L_m f^2} \tag{5.8}$$

Flowchart of zero oscillation variable step size P&O MPPT

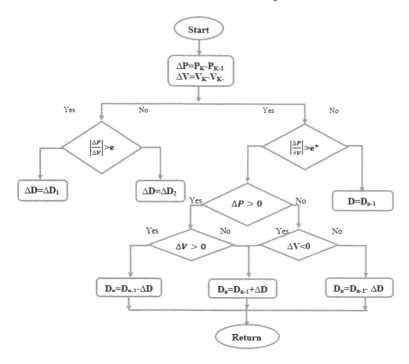

References

1. International Energy Agency, "Global EV outlook 2021 - Accelerating ambitions despite the pandemic", *Global EV Outlook*, vol. 2021, p. 101, 2021. [Online].
2. "Data & Statistics - International Energy Agency", 2021. https://www.iea.org/data-and-statistics.
3. S. F. Tie and C. W. Tan, "A review of energy sources and energy management system in electric vehicles", *Renewable and Sustainable Energy Reviews*, vol. 20, pp. 82–102, 2013.
4. T.-D. Nguyen, S. Li, W. Li, and C. C. Mi, "Feasibility study on bipolar pads for efficient wireless power chargers", *2014 IEEE Applied Power Electronics Conference and Exposition - APEC 2014*, pp. 1676–1682, Mar. 2014, doi: 10.1109/APEC.2014.6803531.
5. E. Akhavan-Rezai, M. F. Shaaban, E. F. El-Saadany, and A. Zidan, "Uncoordinated charging impacts of electric vehicles on electric distribution grids: Normal and fast charging comparison", *2012 IEEE Power and Energy Society General Meeting*, Jul. 2012, pp. 1–7, doi: 10.1109/PESGM.2012.6345583.

134 *Power electronics for electric vehicles*

6. E. A. Nanaki, G. A. Xydis, and C. J. Koroneos, "Electric vehicle deployment in urban areas", *Indoor and Built Environment*, vol. 25, no. 7, pp. 1065–1074, Nov. 2016, doi: 10.1177/1420326X15623078.

7. B. Battke, T. S. Schmidt, D. Grosspietsch, and V. H. Hoffmann, "A review and probabilistic model of lifecycle costs of stationary batteries in multiple applications", *Renewable and Sustainable Energy Reviews*, vol. 25, pp. 240–250, Sep. 2013, doi: 10.1016/j.rser.2013.04.023.

8. D. B. Richardson, "Electric vehicles and the electric grid: A review of modeling approaches, Impacts, and renewable energy integration", *Renewable and Sustainable Energy Reviews*, vol. 19, pp. 247–254, Mar. 2013, doi: 10.1016/j.rser.2012.11.042.

9. P. J. Tulpule, V. Marano, S. Yurkovich, and G. Rizzoni, "Economic and environmental impacts of a PV powered workplace parking garage charging station", *Applied Energy*, vol. 108, pp. 323–332, Aug. 2013, doi: 10.1016/j.apenergy.2013.02.068.

10. D. P. Birnie, "Solar-to-vehicle (S2V) systems for powering commuters of the future", *Journal of Power Sources*, vol. 186, no. 2, pp. 539–542, Jan. 2009, doi: 10.1016/j.jpowsour.2008.09.118.

11. J. Traube, F. Lu, and D. Maksimovic, "Photovoltaic power system with integrated electric vehicle DC charger and enhanced grid support", *2012 15th International Power Electronics and Motion Control Conference (EPE/PEMC)*, pp. LS1d.5-1–LS1d.5-5, Sep. 2012, doi: 10.1109/EPEPEMC.2012.6397399.

12. F. K. Tuffner, M. C. W. Kintner-Meyer, F. S. Chassin, and K. Gowri, "Utilizing electric vehicles to assist integration of large penetrations of distributed photovoltaic generation", 2012.

13. H.-M. Neumann, D. Schär, and F. Baumgartner, "The potential of photovoltaic carports to cover the energy demand of road passenger transport", *Progress in Photovoltaics: Research and Applications*, vol. 20, no. 6, pp. 639–649, Sep. 2012, doi: 10.1002/pip.1199.

14. Hassan Fathabadi, "Novel solar powered electric vehicle charging station with the capability of vehicle-to-grid", *Solar Energy*, vol. 142, pp. 136–143, 2017.

15. Mohammed S. Sheik, D. Devaraj and T. P. Imthias Ahamed, "A novel hybrid maximum power point tracking technique using perturb & observe algorithm and learning automata for solar PV system", *Energy*, vol. 112, pp. 1096–1106, July 2016.

16. Mohammed S. Sheik, D. Devaraj and T. P. Imthias Ahamed, "Genetic optimized fuzzy maximum power point tracking algorithm for solar PV system for abruptly changing weather conditions", *IEI (B) Journal*, vol. 102, no. 3, pp.497–508, 2021.

17. E. P. Sarika, Joseph Kutty Jacob, Mohammed S. Sheik and Shiny Paul, "A novel hybrid maximum power point tracking technique with zero oscillation based on P&O algorithm", *International Journal of Renewable Energy Research*, vol. 10, no. 4, pp. 1962–1973, 2020.

18. Mohammed S. Sheik, "Multiple step size perturb and observe maximum power point tracking algorithm with zero oscillation for solar PV applications", *2018 IEEE International Conference on Current Trends towards Converging Technologies (IEEE ICCTCT 2018)*, Coimbatore, India, March 2018.

19. Mohammed S. Sheik, and J. Ayisha, "Fuzzy logic based energy management of solar powered electric vehicle charging station", *2nd Electric Power and Renewable Energy Conference (EPREC-2021)*, May 2021.
20. Mohammed S. Sheik, and D. Devaraj, "Simulation of incremental conductance MPPT based two phase interleaved boost converter using MATLAB/simulink", *IEEE International Conference on Electrical Computer and Communication Technologies*, pp. 1–6, 2015.
21. Sarah Mary Thomas and Mohammed S. Sheik, "Solar powered EV charging station with G2V and V2G charging configuration", *Journal of Green Engineering*, vol. 10, no. 4, pp. 1704–1731, 2020.

6 Wireless power transfer-based next-generation electric vehicle charging technology

Priyanka Tiwari and Deepak Ronanki

6.1 Introduction

The issues with the widespread use of electric vehicles (EVs) are their slow charging, large battery pack requirements, and the low driving range [1, 2]. The performance of battery systems mainly depends on their charge and discharge profile. Therefore, battery chargers play a vital ;,role in the nurture of EVs, and these are categorized as onboard and offboard chargers [2]. Conductive/wired charging is an extensively used technology to charge EV batteries. However, plugging in and plugging out the charging cable every time can make EV charging complex and unsafe, especially in wet and snowy environments. Furthermore, EVs requires to be equipped with large batteries because they do not support opportunistic charging facility. Also, it requires substantial land on parking lots or the sides of highways, resulting in potential traffic congestion.

Wireless power transfer (WPT)-based EV charging is the most prominent potential solution that can reduce the wires and cables used in the charging system [3]. As a result, the charging of EVs will become more reliable and automatic with lower battery requirements. The salient features of WPT-based (contactless) EV charging are safety, galvanic isolation, weather immunity, flexibility, and automatic charging with fewer maintenance requirements [4–7]. With the help of opportunistic charging, the battery size can also be reduced by up to 20–40% depending on charging level, vehicle type, and road conditions [8]. Furthermore, it offers inherent galvanic isolation between the EV and the grid. WPT-based EV chargers can be classified as stationary, quasi-stationary, and dynamic chargers [9]. Static charging systems are generally suitable for residential charging and public and private parking areas (shopping complexes, corporate offices, etc.) where vehicles are parked for a long time. In quasi-stationary charging, the opportunistic charging can be done in the places where the vehicle waits for a short time, such as traffic signals, bus stops, laybys, and rest areas. Since dynamic charging is an in-motion process, there is no need to wait for complete charging. Therefore, highway lanes and roads must be electrified for dynamic charging applications.

DOI: 10.1201/9781003248484-6

138 *Power electronics for electric vehicles*

The main challenges associated with WPT-based charging systems are efficient design, cost-effective installation, misalignment-tolerant operation, and maintenance requirements. Extensive research has been conducted on power conversion systems and their controls, including coupler design, power electronics converters, materials, communications, and foreign object detection approaches to address these challenges. With these significant improvements, the system efficiency of static chargers has reached 92% at an air gap of 610 mm with a power level ranging from 3.3 to 10 kW [10]. The main objective of this chapter is to provide an overview of the inductive power transfer (IPT)-based EV charging systems due to their popularity and maturity.

Considering the EV charging requirements, this chapter presents the fundamentals of wireless charging technologies, the state of the art, and the design of IPT-based EV chargers. This chapter is organized as follows:

- The fundamentals, requirements, and standards associated with wireless-based EV charging technologies are described in Section 6.2.
- The overview of subsystems such as coils, core structures, control schemes, and power conversion systems pertaining to IPT-based charging systems for EV applications is presented in Section 6.3.
- The implementation of the 3.3 kW IPT charger, including coil design, finite element analysis (FEA), and performance studies, is outlined in Section 6.4. A case study has also been performed in the MATLAB Simulink environment for a 3.3 kW inductive wireless charger to allow EV batteries to be charged with a constant current (CC) / constant voltage (CV) charging profile.
- The future opportunities and advancements of the IPT-based charging systems are presented in Section 6.5.
- Finally, concluding remarks are presented in Section 6.6.

6.2 Fundamentals of wireless power transfer (WPT) technology

WPT refers to the transmission of electrical energy from source to load without a wired connection. In general, WPT can be classified as far-field and near-field. Far-field WPTs can transmit energy through an antenna for long distances in the form of either electromagnetic radiation or mechanical wave radiation. Therefore, they are not considered a good candidate for EV charging applications. In the near field, WPT-based power transfer is done at a distance within the wavelength of the transmitter. Near-field WPT can transmit the power ranging from a few millimetres (mm) to tens of centimetres (cm). Hence, near-field methods are generally preferred for EV charging applications.

6.2.1 Classification of WPT-based battery chargers

The near-field WPT technology for EV charging can be classified into IPT, capacitive power transfer (CPT), and hybrid power transfer (HPT)-based charging systems, as shown in Figure 6.1. The IPT-based technology transfers the power in the form of a magnetic field, whereas CPT transfers power in the form of an electric field. In HPT systems, magnetic and electric fields are used simultaneously to transfer power from the transmitter to the receiver. IPT has been widely used among the near-field WPTs due to its high-power transfer capability at relatively lower frequencies (around 85 kHz).

Based on the type of architecture, WPT-based battery chargers are classified into three categories: static, quasi-static, and dynamic wireless charging systems, shown in Figure 6.1.

6.2.1.1 Static charging of EVs

Static charging refers to the WPT-based EV charging where the receiver is placed aligned with the transmitter to charge the EV, as shown in Figure 6.2. In this configuration, the transmitter side of the coupled coil is placed underground of the parking area, and the receiver side coil is placed at the bottom

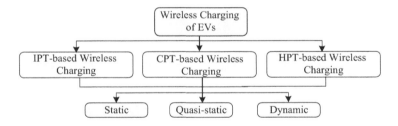

Figure 6.1 Classification of WPT-based EV charging systems.

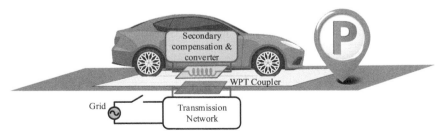

Figure 6.2 Static WPT charger.

140 *Power electronics for electric vehicles*

of the vehicle chassis. Due to the receiver being static, this WPT charging method is simple with easy control. Stationary (static) WPT chargers can be installed in public and private parking areas (shopping complexes, corporate offices), homes, garages, bus stops, etc. Also, it will not obstruct the driver's view, as these chargers are buried on the ground surface. The commercially available static WPT chargers are listed in Table 6.1. Quasi-static charging extends static charging to utilize the opportunity of charging during small-time duration stops of EVs [8]. Hence, they are usually installed at traffic lights, laybys, and rest areas along highways.

6.2.1.2 *Dynamic charging of EV*

Dynamic wireless charging (DWC) is also known as in-motion charging, where the transmitter couplers are installed on a dedicated charging road. Hence, there is no need to wait for the complete battery charge, and it provides inherent galvanic isolation between the vehicle and the grid side [11]. This type of charging can significantly reduce the battery requirement and enhance the driving range by creating frequent charging opportunities. The concept of DWC was suggested by J. G. Bolger in 1978 [12]. The transmitter track is placed under the road, and the receiver is attached to the vehicle chassis. The DWC overcomes the issues with conductive charging, such as range anxiety, charging time, battery size, etc. [13]. Ideally, DWC does not require any battery in the EV. Instead, it can employ supercapacitors, and power from the power rail can continuously charge the EVs [6, 14]. Due to high infrastructure cost, complex control, and lower efficiency in comparison to wired charging, dynamic charging systems have not been greatly commercialized. Some of the commercially available DWC systems are listed in Table 6.2 [10].

Based on the type of transmitter, dynamic chargers can be of four types, [10] as follows:

a) **Segmented track with individual inverters:** The transmitter of these chargers is divided into small segments, and each of the segments is

Table 6.1 Commercially Available Stationary WPT Chargers [5, 7]

Manufacturer	Frequency (kHz)	Airgap (mm)	Maximum power (kW)	Efficiency (%)
WiTricity	145	180	3.3	90
Conductive Wampfler	20	40	60–180	>90
Momentum dynamics	–	610	3.3–10	92
Hevo Power	85	305	1–10	85

Wireless power transfer-based charging 141

Table 6.2 WPT Technologies for Dynamic Charging [10]

Company	Transmitter/ Receiver	Frequency (kKz)	Air gap (mm)	Power (kW)	Efficiency (%)
Korea Advanced Institute of Science and Technology (KAIST)	I/DD	–	240	27	80
Oak Ridge National Laboratory (ORNL)	Rectangular + ferrite bars / Rectangular + ferrite core	22	162	20	93
Bombardier	–	20	60	200	90
Conductix-Wampfler AG	E-Core/F-Core	20	40	120	90
Wireless Advanced Vehicle Electrification (WAVE)	–	20	250	50	90
Utah-State University	Circular + ferrite bars	20	–	25	86

supplied with an individual inverter, which is shown in Figure 6.3a. The process of charging is similar to static chargers with continuously varying mutual inductance. Due to this arrangement, these dynamic charging architectures are most straightforward in design. However, these chargers are difficult to control due to the small segmentations, especially when EVs are operating at high speeds. The efficiencies of short-segmented couplers are around 93% at power levels of 22 kW [10]. One of the major challenges with this configuration is the automatic EV detection system design to activate the powering circuit, which incurs high construction and maintenance costs.

b) **Segmented couplers with distributed switch:** In these dynamic chargers, all segments are connected in parallel with the central supply system, as shown in Figure 6.3b. When a vehicle crosses through a particular segment, that particular switch turns on and transfers power, whereas other segments remain off. This configuration is much more cost-effective than that of individual supplied segments. However, no effective utilization of the central inverter system is observed with this supply configuration [10, 15].

c) **Segmented couplers with centralized power supply:** The track is divided into sub-rails with an individually switched inverter for better energy utilization, as shown in Figure 6.3c. Each sub-rail is connected through n segments and is responsible for supplying power through those segments. This configuration enhances energy utilization and minimizes leakage flux.

d) **Long track transmitter:** A single track is used to transfer the power to a longer distance and hence, there is no effective energy utilization.

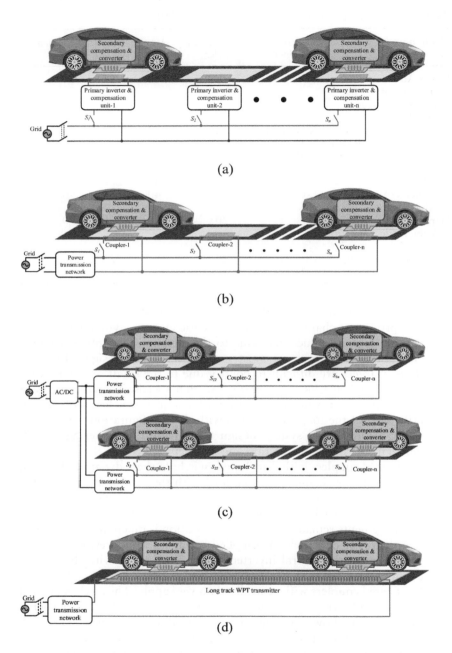

Figure 6.3 Different types of dynamic WPT chargers: (a) short-segmented couplers with individual supply, (b) segmented couplers with distributed switch, (c) segmented couplers with centralized power supply, and (d) long-track WPT transmitter [10].

Wireless power transfer-based charging   143

However, these chargers are easy to control due to their simple architecture. Korea Advanced Institute of Science and Technology (KAIST) initiated this configuration in OnLine EV generations (OLEV, first generation to fifth generation) for dynamic charging. The main advantage of this configuration is misalignment performance can be enhanced to direct the magnetic flux towards the secondary having magnetic material with different core shapes [10, 16].

6.2.2 Standards and requirements

Standardization, regulations, and policies are indispensable in the development and commercialization of reliable and efficient WPT-based charging systems. Hence, several standards and procedures have been progressed in terms of the design, safety, testing and commercialization of these charging systems. Table 6.3 shows the different standards based on the design, testing, efficiency, and safety of WPT-based EV charging systems. Table 6.4 summarizes the WPT class and misalignment requirements provided by the Society of Automotive Engineers (SAE) J2954 [17].

These chargers use electromagnetic induction to transfer power. Hence, the chances of electromagnetic radiation will be high. The radiation can have harmful effects on living beings. Therefore, WPT-based EV charging must follow the guidelines given by the International Commission on Non-Ionizing Radiation Protection (ICNIRP) in 2010. Also, the Institute of Electrical and Electronics Engineers (IEEE) C95.1, 2005 standard has set the limits of maximum permissible exposure of WPT-based charging systems [18].

Therefore, WPT-based chargers should adopt standard approaches and shielding methods, as they emit electromagnetic radiation to the

Table 6.3 Applicable Standards and Codes for WPT-Based EV Charging Systems

Standard	Description
IEC 61980-1, 2, 3	Part-1: General requirement
	Part-2: For communication between WPT-based EV and infrastructure
	Part-3: For the magnetic field power transfer systems
SAE J2954/1	WPT-based light-duty EV / 3–22 kW
SAE J2954/2	WPT-based heavy-duty EV / 22–150 kW
ISO 15118-1, 2, 8	Vehicle to grid communication interface
ISO 19363	Interoperability and safety requirements
IEEE standard C95.1	Restrict the frequency of electric field and magnetic field exposure to outside humans
UL 2750	Wireless charging equipment for EVs
ICNIRP 2010	Guidelines for limiting exposure to time-varying electric and magnetic fields (1–100 kHz)

Table 6.4 WPT Charging Standards and Misalignment Requirements by SAE J2954 [6]

Parameters	WPT power class				Z-class	Gap between GA and VA	Misalignment	
							Offset value (mm)	
Maximum input volt-ampere (VA)	WPT1 3.7 kVA	WPT2 7.7 kVA	WPT3 11.1 kVA	WPT4 22 kVA	Z1	100–150	ΔX	±75
Minimum target efficiency	>85%	>85%	>85%	TBD	Z2	140–210	ΔY	±100
Minimum target efficiency at offset position	>80%	>80%	>80%	TBD	Z3	170–250	ΔZ	$Z_{nom} - \Delta_{low} > Z_{nom} + \Delta_{high}$
Frequency	Preferred frequency 85 kHz In the band of 81.3–90 kHz Roll, rotation, and yaw testing at ±2, ±4, and 6 degrees, respectively							

Table 6.5 Requirements for WPT-Based EV Charger

Parameter	Limit
Efficiency	>85%
Flux density limit B_{max}	<83 A/m
Magnetic field intensity limit H_{max}	<27 µT
Contact current	<17 mA = 2*f (kHz)
Front-end power factor	>0.95
Total harmonic distortion (THD)	<5 %
Ground clearance	160–200 mm
Electric field (E) at frequency range (3–1000 kHz)	614 V/m (general public) 1842 V/m (persons in controlled environments)

surroundings, which will be harmful to living beings. Technical codes and standards essential for the design, safety, testing, and maintenance of WPT-based systems are summarized in Table 6.5.

6.3 Inductive power transfer (IPT)-based EV charging

The basic principle of IPT is based on electromagnetic induction between transmitter and receiver coils, similar to transformers. The power is transferred from transmitter to receiver through loosely coupled coils with a very low coupling coefficient (K). Therefore, a suitable value of a resonating capacitor is connected with both the coils, known as compensating capacitors. These compensating capacitors will increase the maximum power transfer with a minimum volt-ampere (VA) requirement from the supply [19].

Figure 6.4 shows the IPT-based charging scheme for EV applications. It consists of rectification and a power factor correction (PFC) unit to maintain grid codes and regulate direct current (DC) voltage at the nominal value. The PFC stage for WPT chargers is similar to that of conductive/wired chargers, and their details are available in Chapter 1 of this book. The rectified DC is converted to high-frequency AC (HFAC) using a

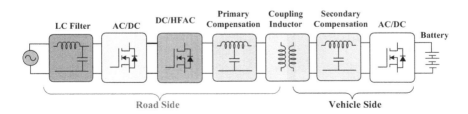

Figure 6.4 Configuration of IPT-based EV charging.

146 *Power electronics for electric vehicles*

transmitter-side HF inverter. This HFAC is supplied to the transmitter coil via the primary compensating unit. The transmitter coil is mutually coupled with the receiver, so the voltage of the same frequency is induced in the receiver coil. The receiver power is fed to the battery via a high-frequency rectification unit at the receiver side. It is mentioned in Chittoor et al. [20] that the maximum load power and efficiency are the functions of frequency. However, switching losses and electromagnetic interference (EMI) issues are observed with higher operating frequencies. Hence, the preferred frequency range for EVs is from 81.3-90 kHz [5]. The main components of IPT-based charging systems are coils, core structures, compensation units, and power electronic converters. These subsystems are explained in detail in other sections of this chapter.

6.3.1 Couplers

Coils are one of the essential parts of an IPT-based charging system. The power transfer, efficiency, voltage gain, coupling coefficient, and stability of IPT systems depend on the transmitter and receiver coil set design [20–22]. The selection of coils is usually based on application requirements such as misalignment, efficiency, transmitter to receiver distance, coupling coefficient values, etc. Hence, selecting an appropriate coil can increase the induced voltage and net power transfer to load. Consequently, the misalignment tolerances can affect the coil design, compensation, converter, and controller design. Therefore, the coil selection and design for static and dynamic charging schemes are different (due to variable mutual inductance in dynamic charging). Generally, planar coils with ferrite bars or sheets are preferred for static charging. Figure 6.5 shows the widely used coils for static charging systems. The design of coils mainly relies on how the magnetic field exists from the transmitter to receiver coils so as to minimize the leakage flux. Table 6.6 compares widely used planar coil structures in terms of their flux orientation, coupling coefficient, and misalignment properties.

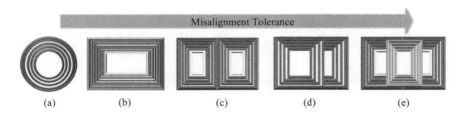

Figure 6.5 Widely used planar coil structures: (a) circular coil, (b) rectangular coil, (c) DD coil, (d) bipolar coil (BP), and (e) DDQ coil [20–22].

Table 6.6 Comparison of Different Planar Coil Structures [16, 22]

Coil Type	Flux Orientation	Coupling Coefficient	Misalignment Tolerance
Circular	Single-sided	Low	Low
Rectangular	Single-sided	Low	Low
DD	Double-sided	Medium	Medium
DDQ	Double-sided	High	High
BP	Double-sided	High	High

The properties of planar coils used for static EV charging are described as follows:

1. **Circular coil set:** These coils are simple in design with the same leakage flux in all directions, i.e., the same misalignment tolerance for all directions. Even though circular coils have a lower coupling coefficient, however, these coils are mostly preferred for static charging applications due to their simple construction.
2. **Rectangular coil set:** The rectangular coil has better forward and backward (lateral) misalignment tolerance than the circular coils due to its rectangular geometry. Furthermore, it exhibits a higher coupling coefficient than circular coils due to lower edge leakage. However, the existence of high inductance at the corners generates higher eddy current losses than in circular coils [20].
3. **DD coil set:** In this configuration, the coil is split into two coils, such that the flux from both the coils should be additive. Therefore, a higher misalignment tolerance with slightly lower output voltage is observed with these coils in comparison to the circular and rectangular coils [23]. Furthermore, these coils require more copper and have a higher cost than circular and rectangular coils due to their complex coil structure [23].
4. **DDQ coil set:** The DDQ coil has two supply coils having 90° phase shifts with each other and supplied by two different inverters. In this type of coil design, lateral and diagonal misalignment tolerance is increased by up to 20% and 35%, respectively [23]. The DDQ coil has the lowest flux leakage with the highest misalignment tolerance. However, these coils demand more copper than all other coil structures.
5. **Bipolar pad (BP):** The structure of the BP coil is shown in Figure 6.5e. It has two identical mutually decoupled coils overlapping at the centre. The BP coil's performance is similar to that of the DDQ coil in terms of misalignment tolerance and coupling coefficient. Furthermore, it requires less copper, thereby offering high power density at a relatively lower cost.

148 Power electronics for electric vehicles

Other coils studied in the literature are tripolar and asymmetric coils, which are used for higher angular misalignment and lateral misalignments, respectively [20]. Figure 6.6 shows the widely used coupler designs in DWC applications, and their properties are listed in Table 6.7.

Table 6.8 shows the comparison of all non-planar couplers in terms of power capacity, efficiency, and distance between the transmitter and receiver coils.

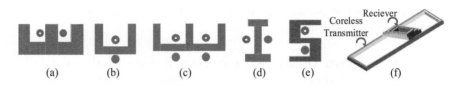

Figure 6.6 Non-planar couplers for IPT-based EV charging: (a) E, (b) U, (c) W, (d) I, (e) ultra-slim S, and (f) coreless [5, 24].

Table 6.7 Widely Used Pick-Up Couplers for DWC [16, 24]

Shape of the Coupler	Properties
E	Simple and low cost
U	These couplers have higher efficiency for a larger air gap
W	Higher power output as compared with U type couplersHave higher coupling coefficient than U type couplersHigh cost
I	Overall cost reduced by 20% compared with W type couplersRelatively lower leakage flux
S	Further reduced cost than I type couplers
Coreless	Due to the absence of core, poor misalignment tolerance observedThe very low coupling coefficientLower output power and efficiency, especially in the cases of dynamic charging applications

Table 6.8 Comparison of Various Non-Planar Couplers [16]

Coupler Shape	Efficiency (%)	Air Gap (cm)	Pick-up Power Capacity (kW)
E	80	1	3
U	72	17	5.2
W	74	20	15
I	80	20	25
S	71	20	22

6.3.2 IPT compensation circuits

In IPT systems, compensation circuits/networks are used to provide compensation for reactive power created by coils. This is done by creating resonance conditions in both the transmitter and receiver coils. An overview and analysis of basic compensation topologies are summarized in Ahmad et al. [25]. The requirements of the compensation circuits are presented as follows:

- Maximize active power at the receiver side
- Minimize supply VA rating
- CV or CC at the output, based on the application requirements
- Maximize efficiency
- Bifurcation-free design
- Highly misalignment tolerant

According to the capacitor connection with the coil, the compensation circuits are classified into four basic types, series-series (SS), series-parallel (SP), parallel-series (PS), and parallel-parallel (PP), as shown in Figure 6.7a [5]. The spider chart shown in Figure 6.7b illustrates the comparison of four basic compensation schemes. It is noticed that compensating capacitor values in SS compensation are independent of load and mutual inductance. Therefore, SS compensation is the preferred compensation technique with a voltage source at the input side [21]. SP compensation should be preferred with the current source due to its high reflected impedance at the input side. PS and PP compensations are generally avoided because compensating capacitance values will depend on coupling coefficient and load. Some other popular higher-order compensation topologies consist of more than one reactive component on each side, such as inductor-capacitor-inductor (LCL), inductor-capacitor-capacitor-inductor (LCCL), and PS capacitor-capacitor [26, 27].

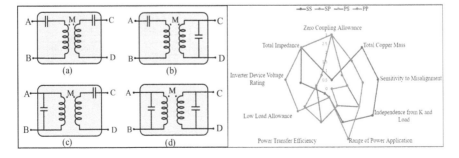

Figure 6.7 (a) Basic compensation circuits, and (b) overall comparison.

6.3.3 Converters and controllers

Power converters play a significant role in maximizing power transfer from the source to the load. Basically, two variants of converters are required, one on the charger side (offboard) and the other on the vehicle side (onboard). The main function of offboard converters is to perform PFC and convert line frequency (50/60 Hz) AC to high-frequency AC (85 kHz), which is supplied to the coupler. Also, it maintains the grid codes (PF >0.95 and total harmonic distortion [THD] <5%) at the input side. On the vehicle side (battery side), the onboard power conversion stage needs to feed power from the receiver coil to charge the battery pack. These onboard or battery-side power converters maintain the power balance and charging specifications demanded by the battery. The classification of offboard and onboard power converters used in IPT-based charging systems is presented in Figure 6.8.

6.3.3.1 Offboard WPT power converters

This section mainly presents an overview of offboard power converters, which are categorized as single-stage and dual-stage based on the number of power electronic stages employed in power conversion from the grid to the transmitter, as shown in Figure 6.9.

The dual-stage configuration comprises two power conversion stages (AC-DC-AC) from line-frequency AC (LFAC) to DC and DC to HFAC, as depicted in Figure 6.9a. A rectifier is used to convert the grid voltage to an intermediate DC-link voltage. Furthermore, it maintains the unity power factor and current THD in compliance with grid codes, which are already defined in Table 6.5. The HFAC inverter connected to DC-link generates a high-frequency current/voltage feeding the primary coil via the primary compensation unit. Based on the application and source used on the primary side, either a voltage source inverter (VSI) or a current source inverter

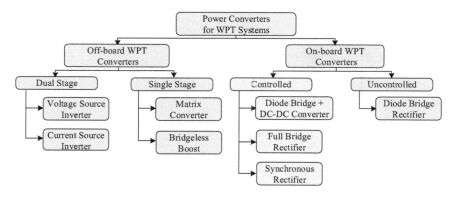

Figure 6.8 Classification of power converters used in WPT-based EV charging.

Wireless power transfer-based charging 151

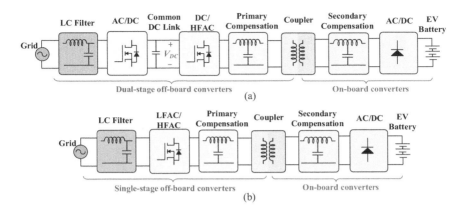

Figure 6.9 (a) Dual-stage IPT-based EV charging scheme, and (b) single-stage IPT-based charging scheme.

(CSI) is employed for the inversion stage. The CSI inverter is preferred with the current source at the input. With CSI, the parallel capacitor compensation on the transmitter side is preferred, such that net reflected impedance to the primary side is maintained high. Similarly, with a voltage source at the input side, series compensation is used at the transmitter side, so that net reflected impedance to the primary side is kept very low. The different variants of dual-stage converters are shown in Figure 6.10a–d [19]. The PFC stage, which is common for conductive and inductive charging, is covered in Chapter 2 of this book. The dominance of the dual-stage power conversion (AC-DC-AC) systems in IPT systems is due to the decoupling nature of the PFC rectifier and the inverter through DC-link capacitors or inductors. Hence, they can be easily controlled to achieve specific or optimized requirements.

On the other hand, direct LFAC–HFAC matrix converters (MCs) are used in single-stage offboard (front-end) converters. This configuration does not demand an immediate DC-link stage, so there is no need for life-limited electrolytic DC-link capacitors, as shown in Figure 6.9b. The MCs can be classified into the buck, half-bridge MC, full-bridge MC, and active clamped H-bridge, as shown in Figure 6.10e–g. However, a double-line frequency ripple appears on the battery side. Hence, sinusoidal ripple current (SRC) charging is the preferred approach with MCs [28]. The performance of widely adopted dual and single-stage power conversion systems is analysed in Huynh et al. [29] to evaluate the best topology in terms of input current THD, power factor, converter power losses, cost, voltage, and current stress. It is claimed that the SRC charging-based single-stage boost-derived matrix converter topology is the preferred option for IPT-based charging systems in terms of efficiency, power density, and cost [29].

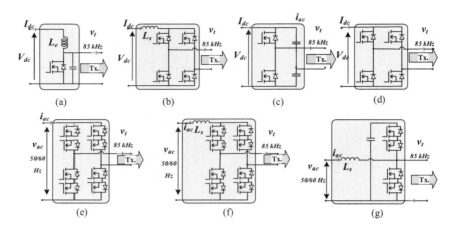

Figure 6.10 Offboard WPT converters for IPT systems: (a) class-E converter, (b) current source inverter, (c) half-bridge voltage source inverter, (d) full-bridge voltage source inverter, (e) full-bridge matrix converter, (f) boost-derived matrix converter, and (g) direct AC–AC active clamped H-bridge [24, 28].

6.3.3.2 Onboard WPT power converters

The possible options on the secondary (battery) side for controllable operations are shown in Figure 6.11a–d. Onboard IPT power converters are selected based on the system control requirements. The diode bridge rectifier is commonly used with static unidirectional charging of EVs, as shown in Figure 6.11a. With single-stage configuration, the issues related to the line current distortion and power factor deterioration are dominant. Hence, controlled converters are required at the battery side to control EV battery charging, as shown in Figure 6.11b and d. With a dual-stage configuration, the

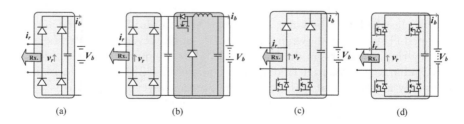

Figure 6.11 Onboard WPT converter for IPT-based charging systems: (a) uncontrolled rectifier, (b) uncontrolled rectifier with DC–DC converter, (c) synchronous rectifier, and (d) controlled bridge rectifier [5, 26, 27].

secondary-side converters are selected based on the type of control scheme used. The synchronous rectifier shown in Figure 6.11c reduces the forward voltage drop due to diodes. A fully controlled H-bridge (HB) is generally preferred in bidirectional power flow cases, as illustrated in Figure 6.11d.

6.3.4 Controllers

The controllers for the WPT system shown in Figure 6.12 can be classified as the primary-side, secondary-side, and dual-side controllers. In the secondary-side control, the battery must have some configuration of controlled converters, as explained in Section 6.3.3.2. This control configuration requires an extra converter in the system. Hence, a charging system with secondary-side control has higher complexity and cost [2]. Based on the transmitter-side HFAC inverter control type, the primary-side control can be classified into three categories: fixed-frequency control, variable frequency control, and dual control. In fixed-frequency control, the system's operating frequency will be fixed at the resonance frequency, and the pulse width of the inverter will be changed by controlling the duty cycle of inverter switches, whereas with variable frequency control, the duty cycle of each switch is fixed at 50%, and the frequency will be varied to achieve the desired control. However, the main issue with the variable frequency control is that the system may go into bifurcation. Hence, the system design will be more complex with variable frequency control. In dual control, both frequency and phase shift are varied to control the desired parameters.

In the primary-side control scheme, all the variables are controlled with offboard, i.e., transmitter-side, converters. Generally, the dual-stage configuration is preferred with the primary-side control scheme. The grid-side converter maintains the grid requirements, and the HFAC inverter controls the EV battery charging. These converters can either get output information from a dedicated communication channel shown in Figure 6.13a or estimate the load parameters from the primary side, as depicted in Figure 6.13b. The

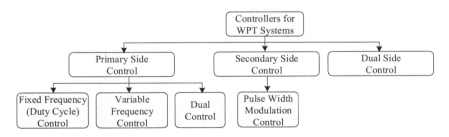

Figure 6.12 Controllers for WPT-based charging systems.

154 Power electronics for electric vehicles

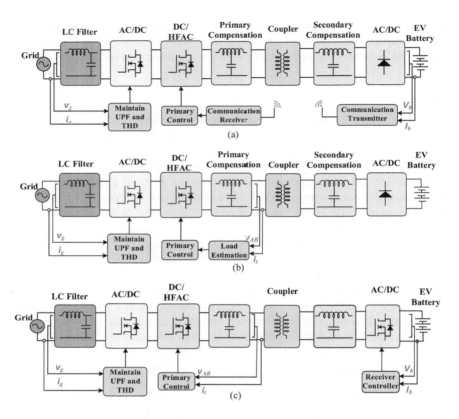

Figure 6.13 Primary side control: (a) with communication, (b) without communication, and (c) dual-side control [28].

control will be more accurate and fast with a communication channel-based primary control scheme. However, the control dependency on the communication link makes the charging unreliable. Primary control without a communication channel will be more reliable but less accurate due to errors in the parameter estimation [30]. Hence, there is a trade-off between accuracy and reliability. Since the system operating conditions are fixed with static charging, primary-side control is generally preferred for static charging of EVs. The battery must be equipped with a controlled rectifier with secondary-side control, as explained in Section 6.3.3.2. Therefore, the secondary-side control is generally preferred, with the systems having multiple receivers to manage power among them. The dual-side controller consists of a fully or semi-controlled rectifier at the battery side, as shown in Figure 6.13c, which is preferred in dynamic charging applications. This can also be applied for bidirectional power flow from vehicle-to-grid (V2G) and grid-to-vehicle (G2V) applications/systems [31, 32].

6.4 Design and simulation of 3.3 kW IPT-based EV charger

This section describes the design and simulation investigation of a 3.3 kW IPT-based EV charger, including the selection of all subsystems in the IPT-based battery charging.

These subsystems consist of a coil set, power converter, compensation circuit, and digital controller. In this section, the first step is to design a coil by considering the battery's power rating. A primary controller for SS compensated IPT system is utilized to control the CC/CV charging of the battery.

6.4.1 Circuital analysis of SS compensated IPT system

Figure 6.14 shows the circuit diagram for a SS compensated WPT system and its equivalent fundamental approximation circuit. The variables C_t and C_r are the transmitter- and receiver-side compensating capacitors, L_t and L_r are the transmitter and receiver coil self inductances, M is the mutual inductance between the coils, and ω_0 is the system resonance frequency. It has been mentioned before that the SS compensation has a capacitance value independent of load and coupling coefficient (K). Furthermore, SS compensation possesses the highest efficiency among all compensation topologies. Hence, an SS compensated IPT system is chosen for the case study. C_t and C_r can be given by:

$$C_t = \frac{1}{\omega_0^2 L_t} \qquad (6.1)$$

$$C_r = \frac{1}{\omega_0^2 L_r} \qquad (6.2)$$

if v_{AB} is the inverter output voltage and v_{CD} is the input voltage to the battery-side rectifier. Applying Kirchhoff's voltage law (KVL) in the transmitter- and receiver-side loop of Figure 6.14 will give current equations as

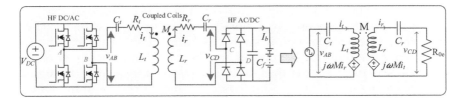

Figure 6.14 Equivalent circuit of SS compensated IPT system.

156   *Power electronics for electric vehicles*

$$i_t = \frac{v_{AB} Z_r}{\omega^2 M^2 + Z_t Z_r} \tag{6.3}$$

$$i_r = \frac{j\omega M v_{AB}}{\omega^2 M^2 + Z_t Z_r} \tag{6.4}$$

At resonance condition,

$$Z_t Z_r = R_t \left(R_r + R_{0e} \right) \tag{6.5}$$

The expression for receiver coil current can be rewritten as

$$i_r = \frac{j\omega M v_{AB}}{\omega^2 M^2 + R_t \left(R_r + R_{0e} \right)} \tag{6.6}$$

where Z_t is transmitter coil impedance, i_t and i_r are the transmitter and receiver coil current, R_t and R_r are the transmitter and receiver coil resistance, R_{0e} is effective EV battery resistance.

The SS compensated IPT system can be represented as a constant current source, which is irrespective of load conditions if the transmitter and receiver coil have negligible resistance, whereas V_{CD} varies according to the load. The transmitter coil current i_t can be written as in Eq. (6.7):

$$\left| i_t \right| = \left| \frac{Z_r i_r}{\omega M} \right| \tag{6.7}$$

From Eq. (6.7), it is observed that as the misalignment between transmitter and receiver coil increases, the current drawn from the source will also increase. Hence, the current drawn from the input side will become infinite at zero coupling. Consequently, the SS compensated IPT system with a voltage source at the input will not allow zero coupling. Thus, it is necessary to put the current limiter on the transmitter side of the IPT system.

6.4.2 Bifurcation analysis

At certain loading and quality factor conditions, it is observed that the system frequency response gives more than one zero phase angle (ZPA) frequency. This phenomenon is known as *bifurcation*. At the bifurcation condition, three resonant frequencies are observed in the phase plot of the system: one of them is lower than the system resonant frequency, the second is greater than the system resonant frequency, and the third is at the system resonant frequency. If the system is in a bifurcation zone, the controlling becomes difficult, especially in the case of variable frequency control. Also, the power transfer capability of the system reduces exponentially if the system is operating in the bifurcation zone. Hence, it is necessary to design the charging system in a bifurcation-free zone. The expressions for the bifurcation-free condition in terms of primary and secondary coil quality factors (Q_p, Q_s) with different compensation circuits are given in Table 6.9.

6.4.3 Coil design

The coil parameters greatly influence efficiency and power transfer in IPT systems [21, 33]. The compensation network is selected based on the system requirements and the self-inductance values of the transmitter and receiver coils. The selection and design of the coils are summarized in two parts:

1. Electrical design of coils
2. Physical design of coils

6.4.3.1 Electrical design of coils

The aim of the electrical design of the coil pair is to find out the inductance values (self and mutual inductance) for both the coils so that they can transfer the required power to load. The load for IPT systems is generally battery, represented as the variable resistance for design purposes. The selection of coupling coefficient K and its relation with voltage gain V_{og}, power transfer to the load P_o, and efficiency η will be analysed in this section.

In the IPT system, it is not always required to increase K to a very high value, as the phenomenon of bifurcation is observed with high coupling and at high load conditions [5, 34]. Its effect may be different for different compensation topologies, as shown in Table 6.9 [31, 35]. Hence, there is a limit on the coupling coefficient (K) to operate IPT chargers in a bifurcation-free zone. The critical coupling coefficient Kc is chosen as 0.25.

Table 6.9 Bifurcation Criterion in Terms of Q_s and K [8]

Compensation topology	Primary capacitance	In terms of Q_p	In terms of K
SS	$\dfrac{1}{\omega^2 L_T}$	$Q_p > 4Q_s^3/\left(4Q_s^2 - 1\right)$	$K < 1/Q_s\sqrt{1 - 1/4Q_s^2}$
SP, PP	$\dfrac{1}{\omega^2 L_T \dfrac{M^2}{L_R}}$	$Q_p > Q_s + 1/Q_s$	$K < 1/\left(Q_s^2 + 1\right)$
PS	$\left(\dfrac{\omega^2 . M^2}{R}\right)^2 + L_T^2\omega^2$	$Q_p > Q_s$	$K < 1/Q_s$

158 *Power electronics for electric vehicles*

6.4.3.2 Electrical parameter estimation

This section includes the estimation of inductance values for both the coils such that they can transfer a given amount of power at the output with a given input rating. The analytical and electrical design of the 3.3 kW charger is done and verified mathematically. The input-output rating of the system is given in Table 6.10. The following steps are to be followed while designing the coil:

STEP 1: It is necessary to estimate the effective equivalent resistance $\left(R_{oe}\right)$ of the battery seen from the receiver coil side. The effective load resistance with a 15% overloading factor for a 3.3 kW battery with battery voltage (V_{o}) 403 V can be estimated in the following way:

$$R_{oe} = \left(\frac{V_{o_{min}}}{I_{o_{max}}}\right) * \left(\frac{8}{\pi^2}\right) * 0.85 = 32\,\Omega \tag{6.8}$$

where

$$V_{o_{min}} = 386.8\,V \text{ and } I_{o_{max}} = 8.32\,A$$

STEP 2: Fix Q_r according to the controlling scheme. High Q indicates low bandwidth, i.e., requires very precise controlling; that is, with a high quality factor, the IPT system becomes difficult to control. In practice, Q_r should be taken between 2 and 10. Here, $Q_r = 4$ is selected.

Table 6.10 Electrical Parameters for 3.3 kW Charger

Parameter	Electrically Designed Values
P_o, P_{in}	3.3 kW
f_o	85 kHz
$V_{0(avg)}$	400 V
V_{dc}	350 V
$L_{t(self)}$	404.22 µH
$L_{r(self)}$	294 µH
K (coupling coefficient)	0.2
K_c (Critical K)	0.248

As we fixed $Q_r = 4$, L_r is obtained as follows:

$$L_r = Q_r \left(R_{oe} + R_r \right) / \left(2\pi f_0 \right) = 294\,\mu H \tag{6.9}$$

where $R_r = 0.45\,\Omega$ and $R_t = 0.91\,\Omega$, resistance of receiver and transmitter coil, respectively.

STEP 3: Mutual inductance (M) between primary and secondary to achieve the desired voltage at the output. At the time of resonance,

$$\left| \omega_0 i_t M \right| = \left| i_r \left(R_{0e} + R_r \right) \right| \tag{6.10}$$

M can be evaluated as follows:

$$M = i_r \left(R_{oe} + R_r \right) / \omega_0 i_t = 68.98\,\mu H \tag{6.11}$$

The input current is assumed to be 10 A for a 3.3 kW, 403 V system (with 8.3 A battery current).

STEP 4: To estimate the value of L_t, the value of the coupling coefficient (K) will be needed. From the bifurcation study, it is clear that there is a limit on K to get a bifurcation-free design. From Table 6.9, the condition on K_c for SS compensation is determined as follows:

$$K < K_c < 1 / Q_r \sqrt{1 - 1 / 4Q_r^2} \Rightarrow K_c = 0.248 \tag{6.12}$$

Hence, the coupling coefficient should not be more than 0.248 for the bifurcation-free operation of an IPT system with the given output voltage and power rating.

STEP 5: After fixing the value of K at 0.2, the value of L_t required is given by

$$L_t = \frac{M^2}{L_r K^2} = 404.22\,\mu H \tag{6.13}$$

From the preceding five steps, it is clear that one can estimate the values of transmitter and receiver coil inductances at the fixed value of coupling coefficient K.

6.4.3.3 Bifurcation verification

To verify the analytical design, the LT-Spice simulation of the SS compensated system is performed with the parameters estimated in the electrical

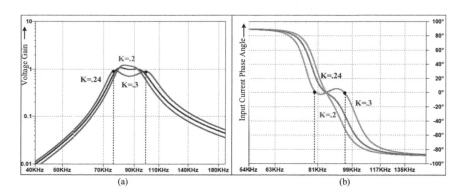

Figure 6.15 Bifurcation-free design validation of coils: (a) phase angle, and (b) voltage magnitude.

design. Figure 6.15a shows that the voltage gain is higher at the lower K values and goes below 1 at $K > K_c$. Hence, at $K > K_c$, the bifurcation occurs, and the voltage gain reduces, reducing the net output power to the receiver. Figure 6.15b shows that when coupling coefficient $K \leq K_c$ (0.2 and 0.24), the transmitter coil current has only one ZPA frequency, which is 85 kHz, whereas in the case when $K > K_c$, i.e., $K = 0.3$, the transmitter coil current has three ZPA frequencies (f_L, f_0, and f_H). This is because of the bifurcation observed at the condition at $K > K_c$.

Figure 6.15b validates the bifurcation-free design of the system at $K = 0.2$. To get high voltage gain and maximum power at the output for a specified load range, the system should not go into the bifurcation zone.

6.4.3.4 Analytical design of coil

The analytical design of the coil includes the estimation of the number of turns required to make the coil designed in Section 6.4.3.3. According to the maximum area available, the outer diameter of the coil has been chosen at 600 mm. While designing the coil analytically, one should consider the following points.

1) **Length of coil:** The length of the coil should be less than the wavelength of the coil at resonance frequency [36]. The wavelength of the copper wire at 85 kHz is in the range of 3.52 km; hence, there is no need to worry about the length of the coil in IPT systems operating at 85 kHz. Furthermore, the effects of the parasitic capacitance of the coils can be ignored if the system resonance frequency is sufficiently lower than the self-resonance frequency (in the range of GHz) of the coil.

   $\omega_0 \ll \omega_{sr}$, where ω_{sr} is the self resonating frequency of the coil,

$$\omega_{sr} = \pi C_0 / l\sqrt{\varepsilon_r \mu_r} = 1 / \sqrt{LC_p} \qquad (6.14)$$

where L, C_p, C_0 are inductance of the coil, parasitic capacitance of coil, and speed of light, respectively; ε_r, μ_r are the relative permittivity and relative permeability of copper.

2) **Minimum strand diameter, skin, and proximity effect:** At high frequency, the resistance of a wire is not as same as of DC or 50 Hz due to the skin effect. Therefore, it is recommended to use the minimum strand of wire, which should be less than 1/4 of skin depth (δ). δ can be defined as the depth in a conductor where current density has decreased to $1/e$ times that of the surface and is given by

$$\delta_{cu|85KHz} = \frac{1}{\sqrt{\pi f \mu \sigma}} = \frac{0.066}{\sqrt{f}} = 0.226 \text{ mm} \qquad (6.15)$$

Where μ is permeability and σ conductivity of coil material. The diameter should be less than 0.056 mm at this frequency minimum strand. Hence, 44 AWG strand wire is selected for making the coils.

3) **Analytic calculation of self-inductance:** The analytical expression to calculate self-inductance with a given number of turns N and fix D_{out} and D_{in} is given by modified Wheeler's expression [21]:

$$L_{self} = N^2 \frac{(D_{out} + D_{in})^2}{8(15D_{out} - 7D_{in})2.54} \mu H \qquad (6.16)$$

This formula can easily give the required number of turns for the inductance values estimated in the electrical design section. For a 3.3 kW charger, the calculated number of turns for the transmitter and the receiver coils is 27 and 19, respectively.

6.4.4 Finite element analysis of the designed coil

The circular spiral coil set with the estimated number of turns in the analytical design has been simulated in the ANSYS Maxwell simulation platform. The results from finite element analysis (FEA) of mutual and self-inductance of analytically designed coils is shown in Figure 6.16. The shape of the transmitter-receiver coil set is shown in Figure 6.16a, and their inductance values under horizontal misalignment are shown in Figure 6.16b. Furthermore, the variation of the coupling coefficient between the transmitter and the receiver coil with respect to the vertical air gap is shown in Figure 6.16c. The comparison of electrical, analytical, and FEA-based coil design is shown in Table 6.11.

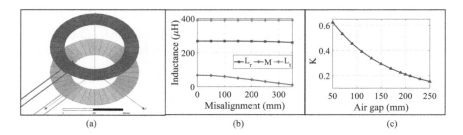

Figure 6.16 ANSYS Maxwell simulation: (a) circular coil design, (b) inductance under horizontal misalignment conditions, and (c) coupling coefficient (K) under different air gaps.

Table 6.11 Comparison of Electrical and Analytical Design of Coils with FEA Simulations

Parameter	Electrical Design	Analytical Design	ANSYS (FEA) Design
L_t	404.22 µH	397.67 µH	398.47 µH
L_r	294 µH	263.3 µH	268.73 µH
M (d = 20 cm)	68.98 µH	–	68.47 µH
K	0.2	–	0.21
N_t	–	27	27
N_r	–	19	19
Conductor Width	–	0.55 mm	0.55 mm
D_{out}	–	600 mm	600 mm

At the final design stage, it is highly recommended to ensure that the designed coil set follows the safety guidelines provided by the ICNIRP 2010. The magnetic field distribution for the designed coil set with 10 A of current through both the coils is plotted in Figure 6.17a and b. This ensures that the safe region starts from 600 mm from the coil center, i.e., flux density lower than 27 µT and magnetic field lower than 83 A/metre.

Table 6.11 compares the electrically designed coil parameter with analytical design and FEA results. From the above table, it is clear that the analytically designed parameters are fairly similar to those of the FEA results, which verifies the accuracy of the designed coil.

6.4.5 Power converter control

CC/CV charging of a battery is safer and more reliable in terms of full charge and life of the battery than CC or CV charging [30]. The primary

Wireless power transfer-based charging 163

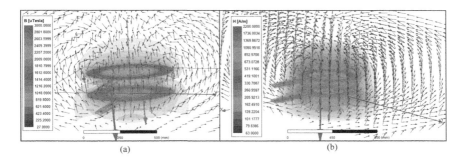

Figure 6.17 ANSYS Maxwell simulation of designed coil set: (a) magnetic flux density, (b) magnetic field strength.

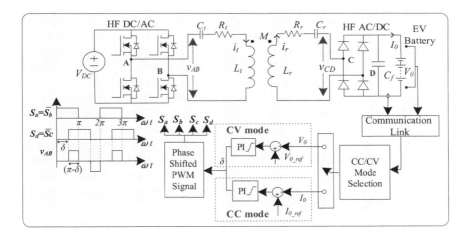

Figure 6.18 Closed-loop control of IPT-based battery charging.

controller has been chosen to control the CC/CV charging of the battery by phase shift control of the HFAC inverter. The phase shift between inverter switches 'a' and 'd' or switches 'b' and 'c' is δ; then, the output voltage of the HFAC inverter v_{AB} can be written as a function of δ, which is $V_{dc}\cos(\delta/2)$. Therefore, the inverter voltage (v_{AB}) can be controlled by just controlling the phase shift angle δ. The gate pulses of HFAC inverter switches S_a, S_b, S_c, and S_d are shown in Figure 6.18. Substituting the (v_{AB}) in Eq. (6.6), the expression for battery charging current can be given as in Tiwari and Tummuru [32]:

$$I_0 = \frac{8}{\pi^2}\left[\frac{\omega M V_{dc}\cos(\delta/2)}{\omega^2 M^2 + R_t(R_r + R_{oe})}\right] \quad (6.17)$$

164   *Power electronics for electric vehicles*

Table 6.12 System Parameters for 3.3 kW IPT Charger

Parameter	Value
DC voltage source, V_{dc}	350 V
WPT operating frequency, f_o	85 kHz
Battery voltage range, V_0	317 to 403 V
Constant charging voltage, $V_{0,max}$	403 V
Constant charging current, $I_{0,max}$	8.32 A
Battery equivalent load resistance, R_o	35–48 Ω in CC 48–400 Ω in CV
Transmitter coil self-inductance, L_t	404.22 μH
Receiver coil self-inductance, L_r	29 μH
Aligned case mutual inductance, M	68.98 μH

From this equation, it is clear that the battery charging current I_o and voltage V_o can be easily controlled by just controlling the phase shift angle δ. All the system parameters are given in Table 6.12. The EV battery is assumed to be equivalent to variable resistance as in [32], and the step variation in load resistance is taken to validate the controller working in CC and CV mode of battery charging. The closed-loop control schematic for IPT-based battery charging systems is depicted in Figure 6.18. The information on battery voltage and current (V_0 and I_0) is collected from the receiver of the communication channel. These charging current and voltage values are given to the mode selection unit. The battery charging mode is selected based on the equivalent battery resistance value in comparison with critical battery resistance (R_{0e_c}). If $R_{0e} < R_{0e\_c}$, the battery will be assumed to be in CC charging mode; otherwise, in CV mode.

At the starting of the battery charging, the battery is assumed to be in CC mode, i.e., the charging current will be at its maximum value $I_{0_{max}}$, and the voltage starts increasing from its minimum value $V_{0_{min}}$. In CC mode, $I_{0_{meas}}$ is compared with the reference value, which is 8.3 A, and the generated error signal is fed to the CC mode proportional and integral (PI) controller, which generates the phase-shifted PWM signal for the HFAC inverter. Once the battery charging voltage crosses its maximum value, i.e. $V_{0_{meas}} \geq V_{0_{max}}$, the battery charging will go into CV mode. In this case, the battery charging voltage is set to its maximum value $V_{0_{max}}$, and then, the current starts reducing. At this time, the error signal ($V_{0_{max}} - V_{0_{meas}}$) is given to the CV mode (PI) controller, which generates PWM pulses for the HFAC inverter. When the charging current reaches 1/10 value, the battery is assumed to be charged.

6.4.6 Case study and discussion

A 3.3 kW IPT charger model is developed and simulated in the MATLAB Simulink platform. The electrical parameters of the 3.3 kW charger are shown in Table 6.12. The CC/CV charging of the EV battery is achieved by phase shift control of the HFAC inverter at the transmitter side, as explained in Section 6.4.4. The voltage profile during the CC mode of charging should rise from its minimum value of 317 V to 403 V. At the same time, the current will be constant at its peak of 8.3 A, i.e., equivalent battery resistance is from approximately 35 to 48 Ω. Similarly, in CV mode, the voltage will be constant at 403 V, and the current starts decreasing from 8.3 A. The battery equivalent resistance ranges from 48 to 480 Ω with the system parameters listed in Table 6.12.

Figure 6.19 shows the inverter output current i_t and voltage v_{AB} waveform for two different coupling coefficient values ($K = 0.2$ and $K = 0.1$). The variation in K emulates the condition of misalignment to validate the controller performance. The completely aligned condition of EV charging is designed at $K = 0.2$, and $K = 0.1$ is taken for the 250 mm misaligned parking condition from the FEA results, as explained in Section 6.4.3. In both the cases, the transmitter-side voltage v_{AB} and current i_t are in phase, as shown in Figure 6.18. This implies that the compensating capacitor requirement is independent of K or M values. Figure 6.19 also concludes that with the increase in misalignment, the current drawn from the source will be increased to maintain the same power at load. Hence, the SS compensated charging system must be designed with the input-side overcurrent protection.

Figure 6.20 depicts the onboard rectifier input voltage v_{CD} and the receiver coil current i_r for both perfectly aligned ($K = 0.2$) and misaligned conditions ($K = 0.1$). From these waveforms, it can be seen that the EV will be charged with the same power level even in the case of misalignments.

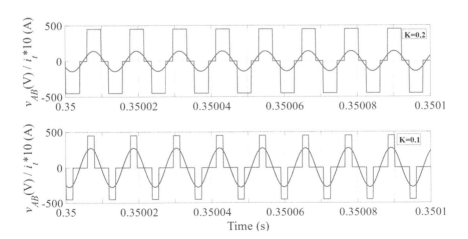

Figure 6.19 Transmitter current and HFAC inverter output voltage @ $M = 68.98$ μH, i.e., $K = 0.2$, @ $M = 35.98$ μH, i.e., $K = 0.1$.

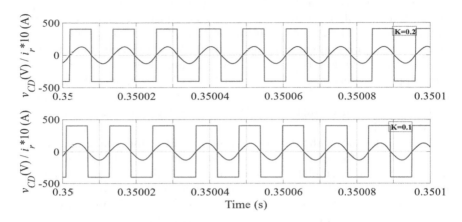

Figure 6.20 Receiver current and HF rectifier input voltage @ M = 68.98 μH, i.e., K = 0.2, and M = 35.98 μH, i.e., K = 0.

The conclusion can be drawn that with the SS compensated system, the load draws the required power even in the case of misalignments.

Figure 6.21 shows the CC/CV charging for both aligned and misaligned conditions. Charging starts at $t = 0$ s,, and the battery voltage is at its minimum value of 317 V, i.e., charging is in CC mode. In CC mode, the battery charging current is at a maximum value of 8.3 A, and the voltage increases from its minimum value. To mimic CC mode, the battery equivalent resistance changes in two steps from 37 to 43 and 46 Ω at time t = 0.1, 0.2 and 0.3 s, respectively. Once battery voltage reaches its maximum value of 403 V at t = 0.4 s, battery charging switches from CC mode to CV mode. In CV mode, now battery voltage will be at its maximum value of 403 V, and the battery charging current starts reducing. In CV mode, the battery equivalent resistance changes in three steps from 80 to 120, 225, and 410 Ω to mimic the CV mode of charging. Hence, the charger works fine for the conditions mentioned earlier and shows its validity in CC/CV charging of the battery. The controller's validity has been observed in the case of misalignment also. From these results, it can be concluded that the designed IPT charger works effectively even when the vehicle is not parked accurately.

6.5 Future trends and opportunities

It is expected that EVs are likely to share 30% of the automobile market by 2030 [37]. However, some major concerns, such as low range, longer charging time, large battery weight, high capital cost, etc., still make hindrances to the global acceptance of EV technology. Wireless charging of EVs is envisioned as a promising solution to circumvent the aforementioned concerns

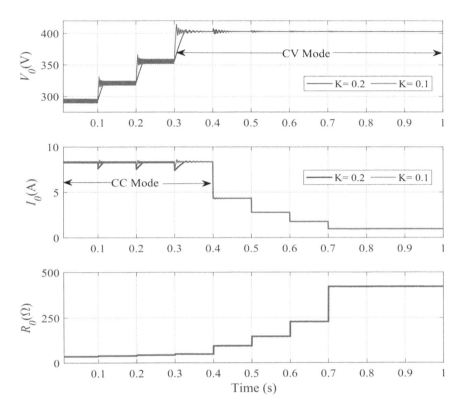

Figure 6.21 CC-CV charging @ $M = 68.98$ µH, i.e., $K = 0.2$, and CC-CV charging @ $M = 35.98$ µH, i.e., $K = 0.1$.

related to EVs, as it is inherently safe during charging and convenient, with weather immunity and the possibility of range extension. However, lower efficiency, low misalignment tolerance, thermal issues, high infrastructural cost, and control and communication are major constraints associated with IPT systems. Over the last two decades, remarkable contributions have been made by researchers to improve the performance of IPT systems through novel coupler design, compensation networks, power converter topologies, and related control techniques. However, there is significant scope for improving IPT systems; the major areas of research opportunities are summarized in Figure 6.22.

Use of Wide-Bandgap Devices: IPT power conversion systems are typically operated at 79–90 kHz as per SAE-J2954 standards. However, operating the power converter at such high frequencies increases switching

Figure 6.22 Future opportunities in wireless charging of EVs.

losses, thereby reducing the efficiency of IPT systems. Utilizing wide-bandgap devices such as silicon carbide (SiC) and gallium nitride (GaN) can reduce the voltage drop across the switch and enable the system to operate at higher frequencies with reduced conduction and switching losses [38]. Additionally, the size of coils and compensation requirements can be reduced at these operating frequencies [39]. Thus, many more research initiatives are needed to improve wide-bandgap devices, especially in device packing, gate driver designs, busbar layout, thermal management, and reliability.

Standardization: The adaptation of EVs made by different manufacturers with the same transmitter could be a big problem in the future and lead to interoperability issues, although standards on power levels are already given in SAE-J2954, and field exposure limits given in ICNIRP-2010 are mostly limited to 22 kW. Therefore, interoperability needs to be defined among various components of IPT systems, including power converter topologies, compensation techniques, charging pads, etc., and their geometries to establish compatibility [40]. There must be some interoperability standards between static and dynamic charging. Furthermore, a generalized coupler design that can support both static and dynamic charging in the same vehicle could be investigated.

Foreign Object Detection: The presence of any metal object within the magnetic field zone near the transmitter of the IPT systems can cause a temperature rise in the metal object. Consequently, many researchers are focusing on foreign object detection (FOD) methods, the most popular being impedance-based metal object detection and temperature rise-based foreign metal object detection. However, it is quite challenging to detect the effects of organic and non-metallic objects, and this requires much more research focus. Hence, advanced FOD techniques need to be developed that can detect metal, non-metal, and organic materials with minimum sensor requirements.

Field Exposure and Safety: Electromagnetic and electric field emissions inevitably become a prime concern for safety, especially in developing high-power WPT systems, and thus demand sophisticated design. Furthermore, shielding design becomes complex during coil misalignment cases due to the further addition of magnetic field emissions. Therefore, it is highly recommended to consider the worst misalignment case in addition to stray field limitations and safety margins. Also, the accuracy in the measurement of the leakage field is questionable at higher power levels, especially in analysing its impacts on the human body. Therefore, accurate measurement of stray fields for accurate shielding design may be a major concern at power levels above 100 kW [41]. Hence, advanced instruments to measure the electric/ magnetic fields and novel shielding methods need to be explored for high-power wireless charging systems.

Grid Stability and Power Quality: The addition of large numbers of wireless chargers, especially dynamic chargers, may cause deterioration of the grid power quality due to their variable power demand. Therefore, it is necessary to focus on the optimized design of the dynamic charging route based on the load profile [42]. Another stimulating area of research could be bidirectional power transfer and renewable integration for wireless charging. The renewable integration can reduce the stress on the utility grid due to a large number of EVs charging simultaneously. The power transfer from V2G can support the utility grid during peak load demand. Hence, more consideration needs to be given to bidirectional power transfer with coordinated control among all the available renewable sources. Accurate power measurement at 85 kHz is a challenging task. The 1% error in power measurement can cause a significant economic loss due to inaccurate billing when the majority of EVs are wirelessly charged [43].

System Installation and Control: IPT systems require installation of the transmitter coils under the road. This requires high mechanical strength of coils, such that they can withstand the weight of the vehicles. Consequently, the roads are constructed with a mixture of magnetic material with cement, termed *magment*. This will increase the strength of magnetic material while reducing infrastructure cost. More research needs to be carried out to investigate the life and cost of road infrastructure, packaging of coils, and material selection for static and dynamic charging configurations.

Lower efficiency and control complexity are the primary concerns for wireless charging. The efficiency of IPT systems can be increased by reducing the power conversion stages. Hence, a single-stage IPT configuration with suitable control can be analysed to increase the efficiency of the charging system. However, it is difficult to control

170 *Power electronics for electric vehicles*

load and grid-side parameters with the single-stage scheme without increasing the number of converters on the EV side. Advanced control techniques such as sliding mode control, model predictive control, and distance observer control can be used to improve the controllability of IPT systems.

Communication Link and Cyber Security: The dependency of EV charging on communication links may lead to cyber security issues and unreliable EV charging [30]. It may be possible for the energy to be captured by unauthorized customers who mimic the legitimate customer and cause economic losses. Additionally, the wireless charging communication protocols can be used to leak the identity of consumers and increase the risks of public charging [41]. Hence, estimation-based primary control and decoupled dual-side control techniques can be explored in this regard. To increase the life of EV batteries, EVs must be charged with a controlled charging profile, such as CC/CV charging and SRC charging. Therefore, more research is needed to develop IPT systems with improved battery charging techniques.

Other aspects: Dynamic charging is a way of charging the EV battery wirelessly while it is in motion. Hence, dynamic charging can be an attractive solution to the existing EV issues, as it significantly minimizes the size of the EV battery. However, the main issues associated with dynamic charging are large output power fluctuations, vehicle detection, and the requirement of secondary-side control [25, 41]. Therefore, further research on coil design and compensation techniques is needed to mitigate the variable power problem. Furthermore, the optimized design of the transmitter segments also needs to be addressed. Coupling and cross-coupling among the different segments can affect the power transfer and hence, need to be analysed properly. In addition, sensorless control of dynamic charging, especially at high speeds, can increase the reliability of EV charging [25]. In dynamic charging, the velocity of the vehicle can affect the power transfer to the EV battery; hence, the controller bandwidth should be sufficiently high to incorporate the high-speed effects. Several coil structures with appropriate compensation algorithms might be investigated in terms of better misalignment tolerance and higher power transfer efficiency.

6.6 Concluding remarks

Cordless charging of EVs can be envisaged as the most convenient and reliable solution with minimum human engagement to conquer the issues associated with EVs. Over the past 30 years, near-field wireless power transfer through inductive power transfer has gained attention from industry as well as academia, and several commercial products and prototypes have been developed. This chapter aims to focus on the fundamentals, standards, and sub-components of inductive power transfer-based charging systems. The technical design considerations in regard to couplers, compensation networks, power converter topologies, control, and communication aspects for both static and dynamic inductive charging systems are explained in detail. In this chapter, a step-by-step design procedure for a 3.3 kW bifurcation-free inductive charger is illustrated and validated through case studies in the ANSYS Maxwell simulation platform. The performance of the designed coil in terms of magnetic field intensity and field exposure limits is studied through FEA in ANSYS software and is compared with analytical calculations using modified Wheeler's formula. The results confirm the effectiveness of the designed circular-shaped coil. The CC/CV charging of EV battery has been validated for different alignment cases through MATLAB simulation results. Finally, the future trends and possible research opportunities in the wireless charging area are briefly discussed. Inductive chargers are expected to serve equally to the existing wired chargers and accelerate the mass market penetration of EVs, which can offer an electrically driven means of transportation.

References

1. D. Ronanki, A. Kelkar, and S. S. Williamson, "Extreme fast charging technology—Prospects to enhance sustainable electric transportation," *Energies*, vol. 12, no. 19, pp. 1–17, 2019, doi: 10.3390/en12193721.
2. M. Yilmaz, and P. T. Krein, "Review of battery charger topologies, charging power levels, and infrastructure for plug-in electric and hybrid vehicles," *IEEE Trans. Power Electron.*, vol. 28, no. 5, pp. 2151–2169, 2013, doi: 10.1109/TPEL.2012.2212917.
3. Z. Zhang, H. Pang, A. Georgiadis, and C. Cecati, "Wireless power transfer - An overview," *IEEE Trans. Ind. Electron.*, vol. 66, no. 2, pp. 1044–1058, 2019, doi: 10.1109/TIE.2018.2835378.
4. S. Li, and C. C. Mi, "Wireless power transfer for electric vehicle applications," *IEEE J. Emerg. Sel. Top. Power Electron.*, vol. 3, no. 1, pp. 4–17, 2015, doi: 10.1109/JESTPE.2014.2319453.
5. D. Patil, M. K. McDonough, J. M. Miller, B. Fahimi, and P. T. Balsara, "Wireless power transfer for vehicular applications: Overview and challenges," *IEEE Trans. Transp. Electrif.*, vol. 4, no. 1, pp. 3–37, 2017, doi: 10.1109/TTE.2017.2780627.
6. D. Ronanki, P. S. Huynh, and S. Williamson, "Power electronics for wireless charging of future electric vehicles," in *Emerging Power Converters for Renewable Energy and Electric Vehicles*, 1st ed., vol. I. Md. Rabiul Islam, Md. R. Shah, Md. H. Ali by CRC Press, 2021 pp. 73–11, doi: 10.1201/9781003058472-3.
7. S. Y. Choi, B. W. Gu, S. Y. Jeong, and C. T. Rim, "Advances in wireless power transfer systems for roadway-powered electric vehicles," *IEEE J. Emerg. Sel. Top. Power Electron.*, vol. 3, no. 1, pp. 18–36, 2015, doi: 10.1109/JESTPE.2014.2343674.
8. H. Karneddi, and D. Ronanki, "Driving range extension of electric city buses using opportunity wireless charging," *IECON Proc. (Industrial Electron. Conf.)*, vol. 2021, 2021, doi: 10.1109/IECON48115.2021.9589475.

172 *Power electronics for electric vehicles*

9. M. Al-Saadi, A. Al-Omari, S. Al-Chlaihawi, A. Al-Gizi, and A. Crăciunescu, "Inductive power transfer for charging the electric vehicle batteries," *EEA - Electroteh. Electron. Autom.*, vol. 66, no. 4, pp. 29–39, 2018.

10. D. Vincent, P. S. Huynh, N. A. Azeez, L. Patnaik, and S. S. Williamson, "Evolution of hybrid inductive and capacitive AC links for wireless EV charging - A comparative overview," *IEEE Trans. Transp. Electrif.*, vol. 5, no. 4, pp. 1060–1077, 2019, doi: 10.1109/TTE.2019.2923883.

11. S. G. Rosu, M. Khalilian, V. Cirimele, and P. Guglielmi, "A dynamic wireless charging system for electric vehicles based on DC/AC converters with SiC MOSFET-IGBT switches and resonant gate-drive," 2016, doi: 10.1109/IECON.2016.7793809.

12. J. G. Bolger, F. A. Kirsten, and L. S. Ng, "Inductive power coupling for an electric highway system," in *28th IEEE Vehicular Technology Conference*, vol. 28, pp. 137–144, 1978, doi: 10.1109/VTC.1978.1622522.

13. D. Patil, J. M. Miller, B. Fahimi, P. T. Balsara, and V. Galigekere, "A coil detection system for dynamic wireless charging of electric vehicle," *IEEE Trans. Transp. Electrif.*, vol. 5, no. 4, pp. 988–1003, 2019, doi: 10.1109/TTE.2019.2905981.

14. K. Li, S. C. Tan, and R. Hui, "Dynamic response and stability margin improvement of wireless power receiver systems via right-half-plane zero elimination," *IEEE Trans. Power Electron.*, vol. 36, no. 10, pp. 11196–11207, 2021, doi: 10.1109/TPEL.2021.3074324.

15. T. Fujita, T. Yasuda, and H. Akagi, "A dynamic wireless power transfer system applicable to a stationary system," *IEEE Trans. Ind. Appl.*, vol. 53, no. 4, pp. 3748–3757, 2017, doi: 10.1109/TIA.2017.2680400.

16. D. Patil, "Dynamic wireless power transfer for electric vehicle," Ph.D. dissertation, EE Dept., Univ. of Texas, Dallas, 2019.

17. V. Cirimele *et al.*, "Uncertainty quantification for SAE J2954 compliant static wireless charge components," *IEEE Access*, vol. 8, pp. 171489–171501, 2020, doi: 10.1109/ACCESS.2020.3025052.

18. IEEE, *IEEE Standard for Safety Levels With Respect to Human Exposure to Radio Frequency Electromagnetic Fields, 3 kHz to 300 GHz*, vol. 2005, no. April. 2006.

19. M. Kosik, R. Fajtl, and J. Lettl, "Analysis of bifurcation in two-coil inductive power transfer," *2017 IEEE 18th Work. Control Model. Power Electron. COMPEL 2017*, 2017, doi: 10.1109/COMPEL.2017.8013324.

20. P. K. Chittoor, B. Chokkalingam, and L. Mihet-Popa, "A review on UAV wireless charging: Fundamentals, applications, charging techniques and standards," *IEEE Access*, vol. 9, pp. 69235–69266, 2021, Institute of Electrical and Electronics Engineers Inc., doi: 10.1109/ACCESS.2021.3077041.

21. K. Aditya, and S. S. Williamson, "Design guidelines to avoid bifurcation in a series-series compensated inductive power transfer system," *IEEE Trans. Ind. Electron.*, vol. 66, no. 5, pp. 3973–3982, 2019, doi: 10.1109/TIE.2018.2851953.

22. S. Wang, D. G. Dorrell, Y. Guo, and M. F. Hsieh, "Inductive charging coupler with assistive coils," *IEEE Trans. Magn.*, vol. 52, no. 7, pp. 1–4, 2016, doi: 10.1109/TMAG.2016.2539340.

23. C. T. Rim, and C. Mi, "Introduction to dynamic charging," in *Wireless Power Transfer for Electric Vehicles and Mobile Devices*, pp. 155–160, 2017.

24. A. Mahesh, B. Chokkalingam, and L. Mihet-Popa, "Inductive wireless power transfer charging for electric vehicles-A review," *IEEE Access*, vol. 9, pp. 137667–137713, 2021, doi: 10.1109/ACCESS.2021.3116678.

25. A. Ahmad, M. S. Alam, and R. Chabaan, "A comprehensive review of wireless charging technologies for electric vehicles," *IEEE Trans. Transp. Electrif.*, vol. 4, no. 1, pp. 38–63, 2017, doi: 10.1109/TTE.2017.2771619.

26. A. A. S. Mohamed, A. A. Marim, and O. A. Mohammed, "Magnetic design considerations of bidirectional inductive wireless power transfer system for EV applications," *IEEE Trans. Magn.*, vol. 53, no. 6, pp. 2–6, 2017, doi: 10.1109/TMAG.2017.2656819.

27. K. Aditya, and S. S. Williamson, "Comparative study on primary side control strategies for series-series compensated inductive power transfer system," *IEEE Int. Symp. Ind. Electron.*, vol. 2016, pp. 811–816, 2016, doi: 10.1109/ISIE.2016.7744994.

28. Y. Tang, Y. Chen, U. K. Madawala, D. J. Thrimawithana, and H. Ma, "A new controller for bidirectional wireless power transfer systems," *IEEE Trans. Power Electron.*, vol. 33, no. 10, pp. 9076–9087, 2018, doi: 10.1109/TPEL.2017.2785365.

29. P. S. Huynh, D. Ronanki, D. Vincent, and S. S. Williamson, "Overview and comparative assessment of single-phase power converter topologies of inductive wireless charging systems," *Energies*, vol. 13, no. 9, 2020, doi: 10.3390/en13092150.

30. K. Song, Z. Li, J. Jiang, and C. Zhu, "Constant current/voltage charging operation for series-series and series-parallel compensated wireless power transfer systems employing primary-side controller," *IEEE Trans. Power Electron.*, vol. 33, no. 9, pp. 8065–8080, 2018, doi: 10.1109/TPEL.2017.2767099.

31. J. Li, and K. Ji, "Frequency splitting research of series-parallel type magnetic coupling resonant wireless power transfer system," *Proc. 13th IEEE Conf. Ind. Electron. Appl. ICIEA 2018*, pp. 2254–2257, 2018, doi: 10.1109/ICIEA.2018.8398085.

32. P. Tiwari, and N. R. Tummuru, "Misalignment tolerant primary controller for series-series compensated static wireless charging of battery," *2019 IEEE Transp. Electrif. Conf. ITEC-India 2019*, 2019, doi: 10.1109/ITEC-India48 457.2019.ITECIndia2019-262.

33. H. Kim *et al.*, "Coil design and measurements of automotive magnetic resonant wireless charging system for high-efficiency and low magnetic field leakage," *IEEE Trans. Microw. Theory Tech.*, vol. 64, no. 2, pp. 383–400, 2016, doi: 10.1109/TMTT.2015.2513394.

34. J. Deng, W. Li, T. D. Nguyen, S. Li, and C. C. Mi, "Compact and efficient bipolar coupler for wireless power chargers: Design and analysis," *IEEE Trans. Power Electron.*, vol. 30, no. 11, pp. 6130–6140, 2015, doi: 10.1109/TPEL.2015.2417115.

35. V. Shevchenko, O. Husev, R. Strzelecki, B. Pakhaliuk, N. Poliakov, and N. Strzelecka, "Compensation topologies in IPT systems: Standards, requirements, classification, analysis, comparison and application," *IEEE Access*, vol. 7, pp. 120559–120580, 2019, doi: 10.1109/ACCESS.2019.2937891.

174 *Power electronics for electric vehicles*

36. H. Vázquez-Leal, A. Gallardo-Del-Angel, R. Castañeda-Sheissa, and F. J. González-Martínez, "The phenomenon of wireless energy transfer: Experiments and philosophy," in *Wireless Power Transfer*, K. Y. Kim, Ed. Rijeka: IntechOpen, 2012.

37. Y. Wang, H. T. Luan, Z. Su, N. Zhang, and A. Benslimane, "A secure and efficient wireless charging scheme for electric vehicles in vehicular energy networks," *IEEE Trans. Veh. Technol.*, vol. 71, no. 2, pp. 1491–1508, 2022, doi: 10.1109/TVT.2021.3131776.

38. V. P. Galigekere *et al.*, "Design and implementation of an optimized 100 kW stationary wireless charging system for EV battery recharging," in *2018 IEEE Energy Conversion Congress and Exposition (ECCE)*, pp. 3587–3592, 2018, doi: 10.1109/ECCE.2018.8557590.

39. S. Li, S. Lu, and C. C. Mi, "Revolution of electric vehicle charging technologies accelerated by wide bandgap devices," *Proc. IEEE*, vol. 109, no. 6, pp. 985–1003, 2021, doi: 10.1109/JPROC.2021.3071977.

40. W. Zhang, J. C. White, A. M. Abraham, and C. C. Mi, "Loosely coupled transformer structure and interoperability study for EV wireless charging systems," *IEEE Trans. Power Electron.*, vol. 30, no. 11, pp. 6356–6367, 2015, doi: 10.1109/TPEL.2015.2433678.

41. H. Feng, R. Tavakoli, O. C. Onar, and Z. Pantic, "Advances in high-power wireless charging systems: Overview and design considerations," *IEEE Trans. Transp. Electrif.*, vol. 6, no. 3, pp. 886–919, 2020, doi: 10.1109/TTE.2020.3012543.

42. W. Wu, Y. Lin, R. Liu, Y. Li, Y. Zhang, and C. Ma, "Online EV charge scheduling based on time-of-use pricing and peak load minimization: Properties and efficient algorithms," *IEEE Trans. Intell. Transp. Syst.*, vol. 23, no. 1, pp. 572–586, 2022, doi: 10.1109/TITS.2020.3014088.

43. S. Y. Chu, and A.-T. Avestruz, "Transfer-power measurement: A non-contact method for fair and accurate metering of wireless power transfer in electric vehicles," in *2017 IEEE 18th Workshop on Control and Modeling for Power Electronics (COMPEL)*, pp. 1–8, 2017, doi: 10.1109/COMPEL.2017.8013344.

7 Asymmetric clamped mode control for output voltage regulation in wireless battery charging system for EV

Dharavath Kishan, Marupuru Vinod, B Dastagiri Reddy, and Ramani Kannan

7.1 Introduction

The popularity of electric vehicles (EVs) has increased during the last decade due to their environmental friendliness, efficiency and reduced noise [1, 2]. The battery pack is a crucial component in these vehicles, and the battery pack performance depends on the charging/ discharging circumstances. Hence, the type of charger plays a key role. The EV batteries can be charged by using on-board or off-board chargers. Generally, these chargers are connected with a plug-in cable. However, these cables are high-gauge cables which may be inconvenient and have potential hazards [3, 4]. These disadvantages can be overcome by a wireless battery charger. In a wireless battery charger, energy can be transferred without any physical connection over a certain range of distance. Wireless chargers are convenient, safe and find uses in automotives, robotics, medical implantable devices, aerospace, etc. [2–4].

Currently, wireless battery charging (WBC) for electric vehicles has gained huge momentum. The WBC or contactless charging for EV is automatic, low maintenance, reliable and safer [1–5]. WBCs for EVs are classified as static charging, quasi-dynamic charging and dynamic charging. The static charging systems can be utilized at specific parking spots and for residential charging. In quasi-dynamic charging, the charging can be done at bus stops, traffic signals and laybys. Dynamic charging (charging while moving or charging in motion) can be installed in highways.

The major problems and challenges associated with WBC are the design of inductive coils, compensation of leakage inductance, battery charging regulation and efficiency. Various inductive coil structures are presented for a variety of applications [6]. Flat spiral circular coils are preferable for EV applications due to their simple construction and misalignment tolerance [7]. The leakage inductance of the coils is high compared with the magnetizing inductance due to the large air-gap between the transmitter and receiver coils for an inductive WBC. To minimize the leakage inductance, appropriate capacitor compensations are used in the WBC charging system,

DOI: 10.1201/9781003248484-7

176 *Power electronics for electric vehicles*

making the converter involved in the WBC work as a resonant converter. Soft switching and easier control of these converters lead to better efficiency and suitability for high-frequency operation compared with non-resonant converters [5–7].

In a WBC system, several methods are present to achieve the constant current or constant voltage mode of charging [6–15]. The common approach in a WBC system is to control the high-frequency inverter at the primary side of the system. Several literature works have described the conventional phase shift control methods to achieve zero voltage (ZVS) or zero current switching (ZCS) for efficiency enhancement and output voltage/current regulation. During the phase shift control, the semiconductor switches to loose zero voltage switching (ZVS) or ZCS, thereby impacting the efficiency of the resonant converter. To address this, an asymmetrical clamped mode control is proposed for the resonant converter in this chapter.

Considering the EV battery requirement in WBC, this book chapter aims to present the design of a suitable control methodology for the regulation of battery charging without compromising on the efficiency of the resonant converter. The chapter is organised as follows:

- The fundamentals of the inductive WBC, block diagram representation and equivalent circuit model of the inductive WBC are described in Section 7.2.
- The mathematical modelling of the proposed asymmetric clamped mode control of the inductive WBC is presented in Section 7.3.
- A case study on the MATLAB/Simulink implementation of the proposed control method for 3.3 kW inductive WBC and simulation results are discussed in Section 7.4.
- The conclusions and the future scope are given in Section 7.5.

7.2 Basics of inductive wireless power transfer (WPT) technology

7.2.1 Basic theory of inductive wireless power transfer (IWPT)

Inductive wireless battery charging works on the principles described by Maxwell's equations. Maxwell's equations are the vital equations of electromagnetism, which contain Gauss's law of electricity, Ampere's law, Faraday's law of electromagnetic induction and Gauss's law of magnetism.

$$\oint E.ds = \frac{Q}{\varepsilon_o} \tag{7.1}$$

$$\oint B.ds = 0 \tag{7.2}$$

Asymmetric clamped mode control for WPT 177

$$\oint E.dl = \frac{-d\phi_B}{dt} \quad (7.3)$$

$$\oint E.dl = \mu o \left(1 + \varepsilon_o \frac{d\phi_B}{dt}\right) \quad (7.4)$$

The first equation (7.1), Gauss's law for electric field, states that the electrical field is related to the distribution of electric charges. The second equation (7.2) (Ampere's Law) states that there are no sinks or sources for magnetic fields, and all field lines are closed loops. The third equation (7.3) (Faraday's law of electromagnetism) relates the electrical field with electrical charges, i.e., the line integral of the electric field around a closed loop is equal to the negative of the rate of change of the magnetic flux density through the area enclosed by the loop. The fourth equation (7.4) (Gauss's law for magnetic field) relates the magnetic field with current and the change in electric flux.

7.2.2 IWPT system equivalent circuit model

The equivalent circuit of the IWPT system is like a conventional transformer. The generalized block diagram of the IWT system is presented in Figure 7.1. The equations of IWPT can be obtained by using the equivalent circuit model as shown in Figure 7.2. It is assumed that under steady-state conditions, the transmitter is excited with sinusoidal voltage and current.

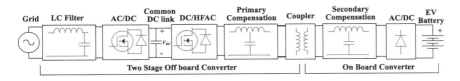

Figure 7.1 Generalized block diagram of the IWPT system.

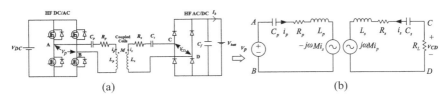

Figure 7.2 (a) DC–DC resonant circuit, (b) fundamental harmonic approximation circuit.

178 *Power electronics for electric vehicles*

By applying Kirchhoff's voltage law (KVL) to Figure 7.2b, the equations attained are as follows

$$Z_P I_P + j\omega M I_S = V_P \tag{7.6}$$

$$j\omega M I_P + Z_S I_S + R_{Leq} I_s = 0 \tag{7.7}$$

where Z_p and Z_s are transmitter and receiver side impedances:

$$Z_P = R_P + j\omega L_P + \frac{1}{j\omega c_p} \tag{7.8}$$

$$Z_S = j\omega L_s + \frac{1}{j\omega c_s} + R_S \tag{7.9}$$

By using Eq. (7.6) in Eq. (7.9), the primary and secondary currents are

$$I_P = \frac{V_P \left[Z_S + R_{Leq} \right]}{Z_P \left[Z_S + R_{Leq} \right] + \omega^2 M^2} \tag{7.10}$$

$$I_S = \frac{-j\omega M I_P}{Z_P \left[Z_S + R_{Leq} \right] + \omega^2 M^2} \tag{7.11}$$

The input and load power can be expressed as follows:

$$P_{in} = \text{Re}\left[V_P \left(I_P \right)^* \right] = \frac{\left(V_P \right)^2 \left[Z_S + R_{Leq} \right]}{\left[Z_S Z_P + R_{Leq} Z_P + \omega^2 M^2 \right]^2} \tag{7.12}$$

$$P_L = \text{Re}\left[\left[I_S \right]^2 R_{Leq} \right] = \frac{\left(V_P \right)^2 \omega^2 M^2 R_{Leq}}{\left[Z_P Z_S + Z_P R_{Leq} + \omega^2 M^2 \right]^2} \tag{7.13}$$

In order to determine the system efficiency, the power loss in wireless coils, semiconductors and system parasitic are neglected.

$$\eta = \frac{P_L}{P_S} = \frac{\omega^2 M^2 * R_{Leq}}{\omega^2 M^2 \left(R_S + R_{Leq} \right) + R_P \left[\left(R_S + R_{Leq} \right)^2 + X_S^2 \right]} \tag{7.14}$$

From Eq. (7.13), the output power can be regulated by controlling the V_P, ω and M. The angular frequency control has many drawbacks, such as a broad noise range, and it has challenges like electromagnetic interference (EMI) issues to monitor. So, controlling ω does not give a feasible solution. Hence, controlling the inverter output voltage (V_p) can control the output power.

In Figure 7.2a, V_p is a high-frequency inverter output voltage, and V_{p1} is the fundamental voltage. The resonant angular frequency and the characteristic impedance of the series compensated converter will be $\omega_0 = \dfrac{1}{\sqrt{L_s C_s}}$, $Z_0 = \sqrt{\dfrac{L_s}{C_s}}$

The switching and normalized switching frequencies are ω_s, $\omega_n = \dfrac{\omega_s}{\omega_0}$ and the quality factor is $Q = \dfrac{Z_0}{R_{eq}}$.

7.3 Control techniques in inductive wireless power transfer system

In IWPT, high power density and efficiency are always the primary goals to achieve. So, resonant converters are best for the IWPT system. With the use of resonant converters, the efficiency of the system increases, and the switching losses decrease by maintaining the soft switching. The main problem with the resonant converters is controlling the switching patterns to get the soft switching. In general, the switching patterns can vary the frequency or duty cycle. The problem with variable frequency controlling the filter design and maintaining the zero voltage switching range is difficult. If you go far beyond the resonant frequency, the conduction losses are increased. Furthermore, a comprehensive working frequency range results in inefficient use of magnetic components and a broad noise spectrum, making it challenging to control EMI.

Many fixed frequency control techniques are proposed to overcome the mentioned difficulties in variable frequency control. Aside from the conventional phase shift control, two more gating schemes, asymmetrical clamped mode control (ACM) [16] and asymmetrical duty cycle control (ADC), are present in the IWPT system. So, the fixed frequency control technique seems to be perfect for maintaining the ZVS. But to get the ZVS, the switching frequency needs to be maintained greater than the resonant frequency. This section describes the control techniques for the series-series resonant converter and analyses their ZVS range through the battery charging profile and under different misalignment conditions. For efficient power transfer, the transmitter and receiver coil should perfectly align; otherwise, the power to the load deviates from the rated conditions, but misalignments always occur, so the mutual inductance varies from the rated conditions. So, suitable closed loop control is required to maintain rated conditions. The dynamic analysis of the series-series resonant converter is performed by using an extended describing function, and the compensator is designed with the help of the k-factor method.

7.3.1 Analysis of phase shift and ACM control strategies

Figure 7.3a and b show the phase shift (PS) and ACM [16] switching patterns and the output voltages of the resonant converter. These control schemes are used to control the output power by maintaining zero voltage across the inverter so the mean voltage is reducing. For the analysis and conceptual understanding, consider that the phase displacement between V_p and V_{p1} is $ø_v$, and the displacement between V_{p1} and I_0 is $ø_1$; therefore, the PS between V_p and I_0 is $Δø$. To achieve ZVS, the $Δø$ has to be greater than zero. In general, to get ZVS, the switching frequency is operated at above the resonant frequency.

In PS control, the first harmonic voltage and phase angle are given by

$$\begin{cases} V_{p1} = \dfrac{4V_{dc}}{\Pi} \cos \dfrac{\alpha}{2} \\ \phi_v = \dfrac{\alpha}{2} \end{cases} \tag{7.15}$$

In the case of the ACM control method, the first harmonic voltage and phase angle are given by

$$V_{P1} = \frac{V_{dc}}{\pi}\sqrt{10 + 6\cos(\alpha)} \tag{7.16}$$

$$\phi_v = \tan^{-1}\frac{\sin\alpha}{3+\cos\alpha} \tag{7.17}$$

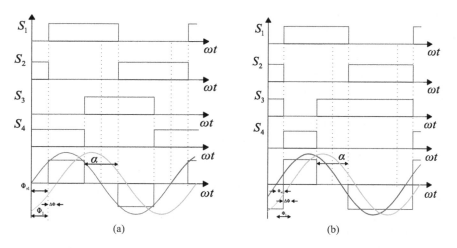

Figure 7.3 Primary side control techniques of IWPT: (a) phase shift, (b) asymmetrical clamped mode control.

The critical value to obtain the ZVS is $\Delta\phi = 0$; from this, we can obtain the frequency in terms of α and Q. The values of ω_n for PS and ACM are given by Eqs (7.18) and (7.19).

$$\omega_n = \frac{\tan\left(\frac{\alpha}{2}\right) + \sqrt{\tan\left(\frac{\alpha}{2}\right)^2 + 4Q^2}}{2Q} \tag{7.18}$$

$$\omega_n = \frac{\sin\alpha(3-\cos\alpha)}{(16+2\sin^2\alpha)Q}$$
$$+ \frac{\sqrt{\left((\cos^2\alpha)-6\cos\alpha+9\right)\sin^2\alpha + \left(256+64\sin^2\alpha+4\sin^4\alpha\right)Q^2}}{(16+2\sin^2\alpha)Q} \tag{7.19}$$

By plotting the graphs of Eqs (7.18) and (7.19), the variation of normalized frequency ω_n function of α and Q can be studied.

It is clear from Figure 7.4, i.e., the normalized switching frequency to get ZVS in PS is increases with the increment in value of α, where as in the ACM control technique the maximum valued of ω_n is 1.06 at Q=3, so the normalized switching frequency to get ZVS is less than that of PS control. Hence, the turn-on losses will reduce, and the system efficiency can be improved.

7.3.2 Mathematical modelling of the proposed control methodology

To design the compensator, the following steps need to followed:

1. Derive the small signal model
2. Get the open loop state space model

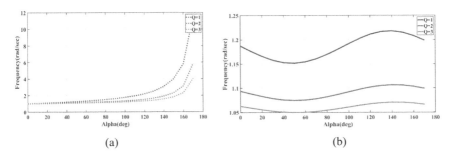

Figure 7.4 Normalized switching frequency to get ZVS with variation of α: (a) PS control, (b) ACM control.

182 *Power electronics for electric vehicles*

3. Find the gain and phase margin with the help of open loop Bode plot
4. Design the compensator with the help of k-factor method

The apparent equivalent resistance between C and D is different from the output resistance because of the nonlinear nature of the switches. So, the resistance between C and D is

$$R_L = \frac{8}{\pi^2} R_0 \qquad (7.20)$$

Let the switching frequency be $\omega_s = \dfrac{1}{\sqrt{L_s C_s}}$

$$C_f \frac{dv_C}{dt} + \frac{v_0}{R_L} = i_s \qquad (7.21)$$

$$\frac{i_s}{i_p} = \frac{\omega M}{R_S + R_L} = n \qquad (7.22)$$

Let the primary side frequency be $\omega_0 = \dfrac{1}{\sqrt{L_p C_p}}$

The switching frequency is selected in such a way that it is greater than the primary side resonance frequency, in order to get ZVS operation. So, transferring the secondary to the primary is shown in Figure .4a.

The secondary side voltage depends on the current; it is positive if the current is positive and negative if the current is negative, as shown in Figure 7.4b.

Apply KVL to the circuit shown in the following:

$$V_{AB} = L_P \frac{di_p}{dt} + V_m + \left(R_P + n^2 R_S\right) i_p + \text{sgn}(i_m) n V_0 \qquad (7.23)$$

$$C_S \frac{dV_m}{dt} = i_p \qquad (7.24)$$

Applying Kirchhoff current law (KCL) to the rectifier network,

$$C_f \frac{dv_0}{dt} + \frac{v_0}{R_L} = i_s \qquad (7.25)$$

The output equation $V_0 = V_{cf}$

7.3.3 Harmonic approximation

The inductor current and the voltage across the capacitor are almost sinusoidal because the resonant tank acts as a bandpass filter. By the use of harmonic approximation, the voltage and current are

$$V_m = V_x \cos(\omega t) + V_y \sin(\omega t) \tag{7.26}$$

$$I_P = I_x \cos(\omega t) + I_y \sin(\omega t) \tag{7.27}$$

By applying the Fourier analysis to Figure 7.5, can obtain the V_{AB}, it has both DC term and harmonics. By considering the fundamental components alone

$$V_{AB} = \frac{V_{DC}}{\pi} \sin(\pi)\cos(\omega t) + \frac{V_{DC}}{\pi}(3 + \cos a)\sin(\omega t) \tag{7.28}$$

So, the voltage across the inverter and the rectifier outputs are nonlinear equations. All other equations are linear, so by the use of the extended describing function mathematical tool, the nonlinear equations are linearized. V_{AB} and V_{CD} can be written as

$$V_{AB} = f_1(d, V_{DC})\sin(\omega t) + f_2(d, V_{DC})\cos(\omega t) \tag{7.29}$$

$$V_{CD} = f_3(I_x, I_Y, V_{cf})\sin(\omega t) + f_4(I_x, I_Y, V_{cf})\cos(\omega t) \tag{7.30}$$

The extended descriptive functions f_1, f_2, f_3, and f_4 in this case establish a connection between the variables that determine the operating point of the converter and the harmonics of the state variables:

$$f_3(I_x, I_Y, V_{cf}) = \frac{4n}{\pi} V_{cf} \frac{I_y}{I_P} \tag{7.31}$$

$$f_4(I_x, I_Y, V_{cf}) = \frac{4n}{\pi} V_{cf} \frac{I_x}{I_P} \tag{7.32}$$

And $i_p = \sqrt{i_x^2 + i_y^2}$

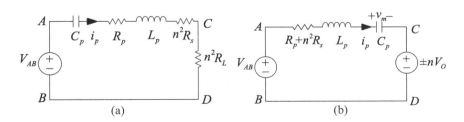

Figure 7.5 Equivalent circuit: (a) secondary referred to primary, (b) final circuit referred to primary.

184 *Power electronics for electric vehicles*

Harmonic Balance:

By equating the DC terms and the sin and cos terms, the equations can be written as

$$L_P \frac{dI_y}{dt} = L_P I_x \omega_s - V_y - \left(R_P + n^2 R_S\right)I_S - \frac{4}{\pi}V_{cf}\frac{I_y}{I_P} + \frac{4V_{DC}}{\pi}\sin\left(3 + \cos\left(2\pi d\right)\right)$$

$$(7.33)$$

$$L_P \frac{dI_x}{dt} = -L_P I_y \omega_s - V_x - \left(R_P + n^2 R_S\right)I_x - \frac{4n}{\pi}V_f\frac{I_x}{I_P} + \frac{V_{DC}}{\pi}\sin 2\pi d \quad (7.34)$$

$$c_p \frac{dv_z}{dt} = I_x - V_y c_p \omega_s \tag{7.35}$$

$$c_p \frac{dv_z}{dt} = I_y + V_y c_p \omega_s \tag{7.36}$$

$$c_f \frac{dv_{cf}}{dt} = \frac{2ni_p}{\pi} - \frac{V_0}{n_0} \tag{7.37}$$

After perturbation and linearization, the state space matrix of the system

$$A = \begin{bmatrix} \frac{-1}{L_P}\left(\frac{4n}{\pi L_P}V_{cf} + R_P + n^2 R_s\right) & -\omega_s & \frac{-1}{L_P} & 0 & \frac{-4nI_x}{\pi I_P L_P} \\[3mm] \omega_s & \frac{-1}{L_P}\left(\frac{4n}{\pi L_P}V_{cf} + R_P + n^2 R_s\right) & 0 & \frac{-1}{L_P} & \frac{-4nI_y}{\pi I_P L_P} \\[3mm] \frac{1}{C_p} & 0 & 0 & -\omega_s & 0 \\[3mm] 0 & \frac{1}{C_p} & \omega_s & 0 & 0 \\[3mm] \frac{nI_p}{\pi I_x C_f} & \frac{nI_p}{\pi I_y C_f} & 0 & 0 & \frac{-1}{R_0 C_f} \end{bmatrix}$$

$$(7.38)$$

$$
B = \begin{bmatrix} \dfrac{-\sin 2\pi d}{\pi L_p} & \dfrac{-2V_{dc}}{L_p}\cos 2\pi d & -I_y \\[2ex] \dfrac{1}{\pi L_p}\left(3+\cos 2\pi d\right) & \dfrac{-2V_{dc}}{L_p}\sin 2\pi d & I_x \\[2ex] 0 & 0 & -V_y \\[1ex] 0 & 0 & V_x \\[1ex] 0 & 0 & 0 \end{bmatrix} \tag{7.39}
$$

$$
C = \begin{bmatrix} 0 & 0 & 0 & 0 & 1 \end{bmatrix} \qquad D = 0
$$

7.4 A case study of 3.3 kW inductive WBC

7.4.1 Analytical modelling

To model the battery charger, we must know the battery's nominal voltage and power rating and at what C-rate the battery is to be charged. In this case, the C-rate and the battery's voltage are considered as 1 and 168 V, respectively. The following steps are involved for the WBC system design.

Step 1

The DC resistance of the battery is computed by Eq. (7.40):

$$
R_L = \frac{8}{\pi^2} R_0 \quad R_0 = \frac{V_0^2}{P_0} \tag{7.40}
$$

The AC equivalent resistance is different from the DC resistance of the battery, so the AC resistance seen by the series-series resonant links is

$$
R_L = \frac{8}{\pi^2} R_0 \tag{7.41}
$$

Step 2

Find the secondary side RMS value of fundamental voltage and current:

$$
v_s = \frac{2\sqrt{2}V_0}{\pi} \tag{7.42}
$$

$$
i_s = \frac{v_s}{R_L} \tag{7.43}
$$

186 *Power electronics for electric vehicles*

Step 3

Let us assume the input power is equal to the output power, and the voltage at the primary side is equal to grid voltage, so the input current is

$$I_p = \frac{v_{prms}}{P_{in}} \tag{7.44}$$

Step 4

Once the values of the primary and secondary current are known, the desired value of mutual inductance can be found to transfer the required amount of power:

$$\left| j\omega_0 M i_p \right| = R_s \left| i_s \right| \tag{7.45}$$

Step 5

The secondary coil inductance is calculated by fixing the secondary side quality factor, so in general, the coil's quality factor is between 2 and 10. In this chapter the quality factor of the coil is considered as 4. If you increase the quality factor of the coil to a higher value, bandwidth will decreases, so the controlling becomes difficult.

$$L_s = \frac{Q_s R_L}{\omega_0} \tag{7.46}$$

Step 6

To avoid the bifurcation, the value of k will be

$$k = \frac{1}{Q_s} \sqrt{1 - \frac{1}{4Q_s^2}} \tag{7.47}$$

Step 7

Once the value of k is calculated, the primary coil inductance can easily be calculated with the help of Eq. (7.48):

$$L_p = \frac{M^2}{L_s k^2} \tag{7.48}$$

7.4.2 Simulation results

The MATLAB Simulink is used to verify the proposed strategy for the 3.3 kW charger. The battery charging profile is created with the help of

Asymmetric clamped mode control for WPT 187

the proposed ACM control. In constant current mode, the rated current is fixed at 19 A, and the battery charging voltage range is varied from 120 to 168 V. Once the battery reaches the nominal voltage, it is the end of the CC mode. In the entire CC mode, the battery's internal resistance changes from 6.2 to 8.9 Ω. In the CV mode, the voltage is kept constant at 168 V, and the currents start decreasing from the rated value to 1% of the rated value, and in the CV mode, the battery equivalent resistance changes from 9 to 500 Ω (Figure 7.6, Table 7.1).

Figure 7.7 shows the simulation results of the inverter output voltage and current at α = 30° and α = 100°, respectively. Similarly, the output voltage and output current at the CC mode and CV mode are prersented in Figure 7.8 and Figure 7.9. The constant current mode will be operated till the battery nominal voltage reaches 168 V. After completion of the CC mode, the CV mode will be activated, and once the charging current reaches 1% of the rated current, the charging process will be terminated.

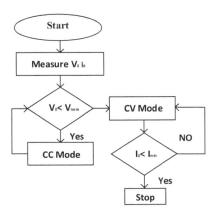

Figure 7.6 Battery charging profile algorithm.

Table 7.1 Electrical Parameters of 3.3 kW System

Parameter	Value
DC input voltage V_{dc}	340 V
Output voltage V_0	168 V
Primary coil inductance L_p	329.8 µH
Secondary coil inductance L_s	51.923 µH
Mutual inductance M	26.175 µH
Resonant frequency f_o	85 kHz
Switching frequency f_s	88 kHz
Load resistance R_0	8.9 Ω

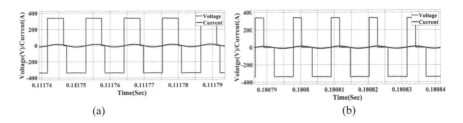

Figure 7.7 Inverter output voltage and current at (a) α = 30° and (b) α = 100°.

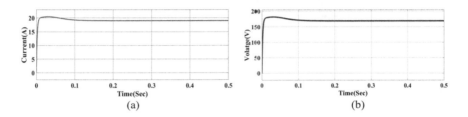

Figure 7.8 CC mode: (a) current, (b) voltage.

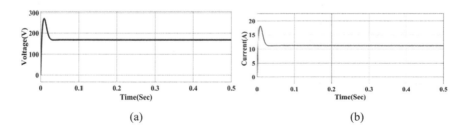

Figure 7.9 CV mode: (a) voltage, (b) current.

7.4.3 Design of closed loop compensator

Figure 7.10 shows the Bode plots of the open loop and closed loop systems. In the open loop system, the gain margin and phase margin are 17.5 db and 14°; even though both are positive, the phase margin is lower. If any disturbance occurs, the system will becomes unstable. So, in general, for a stable system, the phase margin should be between 60° and 90°. Hence, the compensator is designed for an 85° phase margin, and the closed loop Bode plot shown in Figure 7.10b with the desired phase margin.

Asymmetric clamped mode control for WPT 189

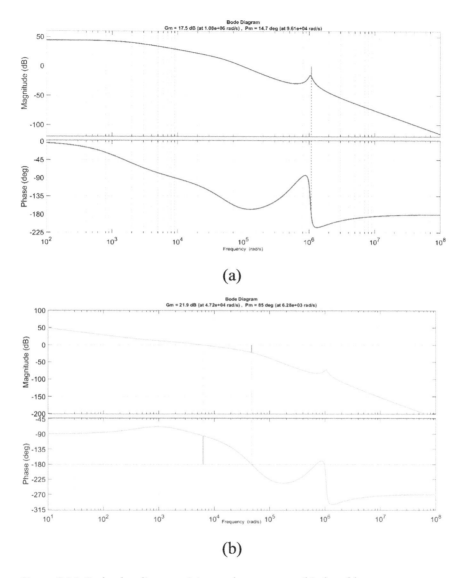

Figure 7.10 Bode plot diagram: (a) open loop system, (b) closed loop system.

Figure 7.11 and Figure 7.12 show the step change in the reference current and load resistances. While carrying the step change in the reference current, the load resistance and mutual inductance are kept at constant values. In step change in load, the reference current and mutual inductance are operated at constant. In both cases, the controller tracks the output regulation with in the specified limits of over shoot.

190 *Power electronics for electric vehicles*

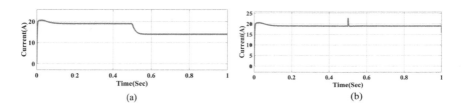

Figure 7.11 Step change in CC mode: (a) reference change, (b) load change.

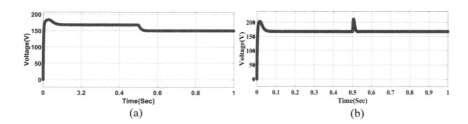

Figure 7.12 Step change in CV mode: (a) reference voltage, (b) load.

7.5 Conclusion

Wireless charging of EVs can be a most convenient and reliable solution to overcome the issues associated with plug-in/conductive EVs charging. This chapter aims to focus on the fundamentals of inductive wireless power transfer-based charging systems. The technical design considerations in control of output voltage and current for the inductive wireless charging system are explained in detail. The mathematical modelling of the closed loop compensator design and its validation for the battery charging are explained. In this chapter, a case study for a 3.3 kW wireless inductive charger is illustrated and validated through the MATLAB simulation platform. The performance of the designed asymmetric clamped mode control for the CC/CV charging of EV battery has been validated for different dynamic conditions.

References

1. A. Emadi, Y. J. Lee and K. Rajashekara, "Power Electronics and Motor Drives in Electric, Hybrid Electric, and Plug-In Hybrid Electric Vehicles," in *IEEE Transactions on Industrial Electronics*, vol. 55, no. 6, pp. 2237–2245, June 2008.
2. R. Bosshard and J. W. Kolar, "Inductive Power Transfer for Electric Vehicle Charging: Technical Challenges and Tradeoffs," in *IEEE Power Electronics Magazine*, vol. 3, no. 3, pp. 22–30, Sept. 2016.

3. D. Kishan and P. S. R. Nayak, "Wireless Power Transfer Technologies for Electric Vehicle Battery Charging — A State of the Art," in *2016 International Conference on Signal Processing, Communication, Power and Embedded System (SCOPES)*, Paralakhemundi, India, 2016, pp. 2069–2073.
4. L. Zhao, D. J. Thrimawithana and U. K. Madawala, "Hybrid Bidirectional Wireless EV Charging System Tolerant to Pad Misalignment," in *IEEE Transactions on Industrial Electronics*, vol. 64, no. 9, pp. 7079–7086, Sept. 2017.
5. W. Zhong and S. Y. R. Hui, "Charging Time Control of Wireless Power Transfer Systems Without Using Mutual Coupling Information and Wireless Communication System," in *IEEE Transactions on Industrial Electronics*, vol. 64, no. 1, pp. 228–235, Jan. 2017.
6. Y. Jiang, L. Wang, J. Fang, R. Li, R. Han and Y. Wang, "A High-Efficiency ZVS Wireless Power Transfer System for Electric Vehicle Charging With Variable Angle Phase Shift Control," in *IEEE Journal of Emerging and Selected Topics in Power Electronics*, vol. 9, no. 2, pp. 2356–2372, April 2021.
7. J. Wu, L. Bie, W. Kong, P. Gao and Y. Wang, "Multi-Frequency Multi-Amplitude Superposition Modulation Method With Phase Shift Optimization for Single Inverter of Wireless Power Transfer System," in *IEEE Transactions on Circuits and Systems I: Regular Papers*, vol. 68, no. 5, pp. 2271–2279, May 2021.
8. Y. D. Lee, D. M. Kim, C. E. Kim and G.-W. Moon, "A New Receiver-Side Integrated Regulator with Phase Shift Control Strategy For Wireless Power Transfer System," in *2020 IEEE PELS Workshop on Emerging Technologies: Wireless Power Transfer (WoW)*, Seoul, Korea (South), pp. 112–115, 2020.
9. Y. Li, J. Hu, F. Chen, Z. Li, Z. He and R. Mai, "Dual-Phase-Shift Control Scheme With Current-Stress and Efficiency Optimization for Wireless Power Transfer Systems," in *IEEE Transactions on Circuits and Systems I: Regular Papers*, vol. 65, no. 9, pp. 3110–3121, Sept. 2018.
10. F. Liu, K. Chen, Z. Zhao, K. Li and L. Yuan, "Transmitter-Side Control of Both the CC and CV Modes for the Wireless EV Charging System With the Weak Communication," in *IEEE Journal of Emerging and Selected Topics in Power Electronics*, vol. 6, no. 2, pp. 955–965, June 2018.
11. Q. Deng, J. Liu, D. Czarkowski, M. Bojarski, E. Asa and F. de Leon, "Design of a Wireless Charging System with a Phase-Controlled Inverter Under Varying Parameters," in *IET Power Electronics*, vol. 9, no. 13, pp. 2461–2470, 2016.
12. Y. Li, J. Hu, F. Chen, Z. Li, Z. He and R. Mai, "Dual-Phase-Shift Control Scheme with Current-Stress and Efficiency Optimization for Wireless Power Transfer Systems," in *IEEE Transactions on Circuits and Systems I: Regular Papers*, vol. 65, no. 9, pp. 3110–3121, Sept. 2018.
13. A. Berger, M. Agostinelli, S. Vesti, J. A. Oliver, J. A. Cobos and M. Huemer, "A Wireless Charging System Applying Phase-Shift and Amplitude Control to Maximize Efficiency and Extractable Power," in *IEEE Transactions on Power Electronics*, vol. 30, no. 11, pp. 6338–6348, Nov. 2015.
14. M. Bojarski, E. Asa, K. Colak and D. Czarkowski, "Analysis and Control of Multiphase Inductively Coupled Resonant Converter for Wireless Electric Vehicle Charger Applications," in *IEEE Transactions on Transportation Electrification*, vol. 3, no. 2, pp. 312–320, June 2017.

192 *Power electronics for electric vehicles*

15. Z. Li, K. Song, J. Jiang and C. Zhu, "Constant Current Charging and Maximum Efficiency Tracking Control Scheme for Supercapacitor Wireless Charging," in *IEEE Transactions on Power Electronics*, vol. 33, no. 10, pp. 9088–9100, Oct. 2018.

16. K. Aditya and S. S. Williamson, "Design Guidelines to Avoid Bifurcation in a Series–Series Compensated Inductive Power Transfer System," in *IEEE Transactions on Industrial Electronics*, vol. 66, no. 5, pp. 3973–3982, May 2019.

8 Selection of electric drive for EVs with emphasis on switched reluctance motor

Pittam Krishna Reddy, P Parthiban, and R Kalpana

8.1 Introduction

The significant emission of greenhouse gas due to the use of internal combustion (IC) engines in the transportation sector can be reduced by replacing IC engines with highly efficient electric vehicles (EVs). Induction motors (IMs), permanent magnet (PM) motors and switched reluctance motors (SRM) are the options for EVs. IMs are popular for EVs due to matured technology, significant research developments and the capability of a digital processor for implementing complex algorithms. PM motors are competitors for IMs due to the use of high-power density rare-earth magnets and relatively easier control; on the other hand, manufacturing complexity and the cost of rare-earth magnets are the drawbacks of these motors [1]. PM motors are categorized as sinusoidal type and trapezoidal type on the basis of air gap flux distribution. Concentrated three-phase windings are placed in trapezoidal flux distribution motors, commonly known as brushless direct current (BLDC) motors. Sinusoidal flux distribution machines, usually called permanent magnet synchronous motors (PMSMs), provide a uniformly distributed rotating magnetic field. Inherent fault-tolerant ability and lack of windings/PMs on the rotor circuit makes the SRM a potential candidate for traction application with improved inertia rating of the motor. The main components of the EVs shown in Figure 8.1 are controller, converter, energy storage system, motor and transmission.

8.2 Motor selection for EV

8.2.1 Electric motor requirements for EV applications

The selection of the motor drive for EV application depends on the fulfilment of cost and power density requirements. The reduced cost of the EV ensures large-scale usage, and higher power density is required to achieve a lighter-weight package. The motor design for EV requires expertise in various aspects, including electromagnetic, thermal, mechanical and structural systems. Therefore, a multidisciplinary approach is adopted when designing motors for EV applications [2]. The performance of the drive and

DOI: 10.1201/9781003248484-8

194 *Power electronics for electric vehicles*

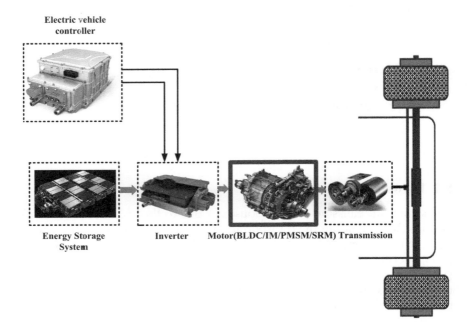

Figure 8.1 Components of EV.

manufacturing cost are decided by the material used and the manufacturing technique adopted.

1. *Torque generation and current density*: Higher current density and higher air gap flux density are needed to achieve higher torque. Improved installation and higher fill factor are required to maintain high current density. However, higher current density results in increased stator copper losses. The stator copper losses can be minimized by reducing winding resistance with the help of proper design of windings and stator core.
2. *Electrical steel*: Electrical steel is used to achieve a higher magnetic flux density, thereby lower current is sufficient to generate the required torque. The finite electric conductivity of the steel results in eddy current losses when exposed to alternating electromagnetic fields. Usually, silicon is added to the steel during the rolling process to reduce the eddy current losses by increasing the resistivity. However, a higher silicon content makes the rolling process harder by introducing brittleness and also reduces the magnetic flux density. Therefore, the silicon content should be limited to 3.5%. Electrical steel sheets are laminated to reduce eddy current losses. Optimized selection of the grade and

Selection of electric drive for EVs 195

thickness of lamination material can reduce the losses and provides higher flex density. A lamination thickness of 0.2 to 0.35 mm is used in traction motors to reduce eddy current losses.

3. *Thermal management system:* Due to the requirement for higher values of torque density, power density and frequency result in higher heat dissipation. Therefore, optimized machine design is required to reduce the losses. Longer operation life can be achieved with low temperature rise. A proper heat dissipation mechanism is required for traction motors to meet reliability, performance and lifetime requirements. The thermal resistance and adopted heat transfer method decide the temperature rise of the machine. The coercivity of the rare-earth magnets is affected by the rise in temperature, which brings significant challenges to the design of PM machines.

4. *Permanent magnets:* PM traction motors have higher torque density and better efficiency due to use of rare-earth magnets. The increase in temperature reduces the coercive force of the PMs. Therefore, extra caution is required when designing PM motors. Coercive force shows the the resistance of the magnet towards demagnetization when an external magnetic field is applied. High temperature reduces the coercive force. Therefore, the behaviour of the magnetic circuit during field reversal has to be studied carefully under high operating temperatures. However, heavy rare-earth elements such as dysprosium are added to increase the intrinsic coercivity.

8.2.2 Speed-torque characteristics

The EV performance is decided by the speed-torque characteristics.

BLDC: The speed-torque characteristics of a BLDC motor are shown in Figure 8.2. The BLDC motor operates in the constant torque region up to the rated speed. If the motor is operating above the rated speed, equal to nearly 150% of rated speed, then the nature of the torque drops. A BLDC motor requires higher than rated torque for a small period during acceleration or starting from standstill; this extra torque is used to overcome the inertia of the mechanical load.

PMSM: The speed-torque characteristics of a PMSM are shown in Figure 8.3. The PMSM operates in the constant torque region up to base speed ω_b and maintains constant power mode above the base speed. If the PMSM is operated above the critical base speed ω_{1b}, the torque will decrease at a faster rate with an increase in speed. Therefore, operating the PMSM above the critical base speed is not feasible. Usually, the PMSM is operated in the constant torque and constant power region.

IM: The speed-torque characteristics of an IM are shown in Figure 8.4. These characteristics are divided into two regions: stable and unstable

196 *Power electronics for electric vehicles*

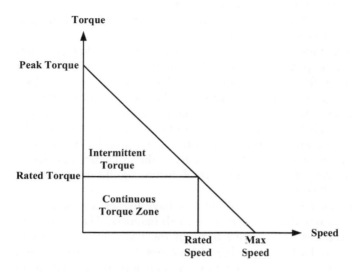

Figure 8.2 Speed-torque characteristics of BLDC.

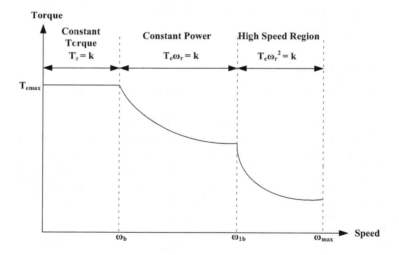

Figure 8.3 Speed-torque characteristics of PMSM.

regions. At a low value of slip, the torque is directly proportional to the slip, i.e., the linear region ($0 < S < S_m$). At a high slip, greater than S_m, the torque and speed are inversely proportional, i.e., the non-linear region ($S_m < S < 1$). The speed range of the IM can be two times the base speed, beyond which breakdown torque will be reached.

Selection of electric drive for EVs 197

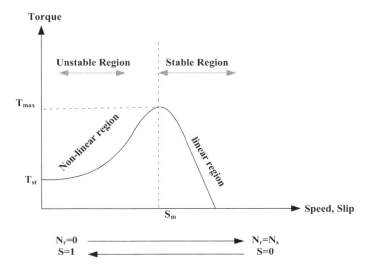

Figure 8.4 Speed-torque characteristics of IM.

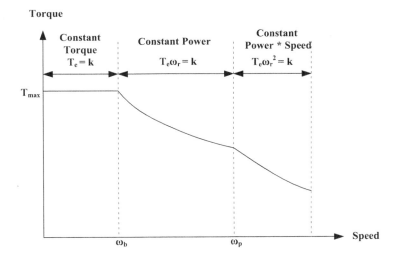

Figure 8.5 Speed-torque characteristics of SRM.

SRM: The speed-torque characteristics of an SRM are presented in Figure 8.5. At low speed, the phase current rises sharply with the excitation due to low back-electromotive force (EMF). Usually, hysteresis current control or voltage pulse width modulation control is used to control the current in an SRM. The back-EMF increases along with the speed, which suppresses the current rise due to fixed excitation voltage at higher speed. So, advanced phase turn-on is required to maintain the desired current level for maximum torque production. The SRM operates in the constant torque region for speed range up to base speed ω_b [3].

Beyond the base speed, if back-EMF exceeds the supply voltage, chopping control or pulse width modulation (PWM) control is not possible because the current starts decreasing once pole overlap begins; therefore, the SRM operates in the single pulse mode at higher speeds. The torque and speed are inversely proportional above the base speed. The decrease in current can be encountered by extending the conduction angle; thereby, the required current is maintained in phase. The constant power operation can be allowed up to the maximum possible conduction angle at speed ω_p. At higher speeds, above ω_p, no further increase of the conduction angle is possible; therefore, torque falls rapidly, as it is inversely proportional to $1/\omega^2$.

8.2.3 Power electronic circuit associated with traction motors

Except for SRM, a standard voltage source inverter circuit is used to drive BLDC, PMSM and IM, as shown in Figure 8.6.

8.2.4 Regenerative braking

The driving motor of the EV can recover the kinetic energy during normal braking. The recovered braking energy is stored in the battery and used as drive power to extend the driving distance of the vehicle. The power electronic switches in the circuit associated with the traction motor will assist in bringing back the kinetic energy to the battery side during braking. And if

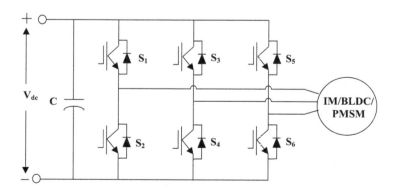

Figure 8.6 Power electronic converter for IM/BLDC/PMSM.

the battery is fully charged, then a resistive load should be used for energy dissipation during braking.

BLDC: Regenerative braking is possible with the same power circuit used to drive the motor. The motor inductance along with the power circuit will be used as a boost circuit to return the energy to the battery during low-speed operation. This is achieved by controlling the H-bridge with PWM.

PMSM: This motor is controlled by the inverter; when braking is applied to the motor, the stored kinetic energy is fed back to the source using proper commutation of switches with the help of feedback diodes.

IM: Regeneration is possible by changing the supply frequency.

8.2.5 Comparison of permanent magnet motors (PMSM/BLDC), IM and SRM

Several motors have been investigated to check their suitability for EV applications. BLDC and PMSM are the preferred choice because of high power density and efficiency due to the use of PMs. However, limited resources of PMs and supply chain issues are leading research towards finding motor drives without PMs, such as IMs, SRMs and synchronous reluctance (SyncRels) machines. Out of these, SRM is found to be the most suitable candidate for traction applications, with its distinctive features of robust and low-cost construction, minimum maintenance, wide constant power region, inherent fault tolerance, and capability of running at high temperature and harsh environmental conditions. Moreover, the rotor doesn't have windings/PMs, which makes high-speed operation possible. However, careful design and optimized control algorithms are required to mitigate torque ripple, noise and vibrations for high-performance applications. The motors most often used for EV applications are compared in Table 8.1.

8.3 Switched reluctance motor

The SRM is gaining more popularity over the conventional alternating current (AC) motor for industrial, residential and automotive applications due to its rugged and low-cost construction, requirement for minimum maintenance, broad constant power region and high efficiency. However, the SRM exhibits non-linear magnetic behaviour due to the salient pole structure of both stator and rotor. SRM phases are excited in succession, due to which torque is produced in a discrete manner. These two factors inherently produce torque ripple, which leads to noise and vibration. The aforementioned concerns can be addressed through motor design improvements and/ or development of sophisticated control techniques. However, the crucial part of the control phenomenon is the proper commutation of the phases to avoid negative torque generation.

200 Power electronics for electric vehicles

Table 8.1 Comparison of Different Motors for EV Applications

Parameter	Permanent Magnet Motor	IM	SRM
Manufacturing Cost	High due to rare-earth magnets	Medium due to rotor windings	Low due to lack of windings/PMs on rotor
Maintenance	High	Medium	Low
Efficiency	High due to presence of PMs	Medium due to rotor copper losses	Low at lower speeds, high at medium and high speeds
Power Density	High	Medium	Medium
Fault Tolerance Capability	Low	Low	High
Torque Ripple	Low	Low	High
Noise and Vibration	Low	Medium	High
Operating Temperature	Low	Medium	High

The significant control algorithms proposed for SRM control are broadly classified into indirect and direct torque control methods. In the indirect torque control method, current profiling and commutation angle optimization techniques are widely used. Current profiling can be achieved through different torque sharing functions (TSFs), which are proposed to appropriately distribute the torque between the phases to achieve the primary objective of torque ripple minimization, especially during the commutation region. However, these control approaches require storing the non-linear magnetic characteristics, which are undesirable for practical implementation. Moreover, extensive computations need to be performed for angle optimization. In contrast to indirect torque control methods, the DTC approach controls the instantaneous torque, with the primary objective of minimizing the torque ripple and enhancing the torque/ampere (T/A) ratio. DTC methods comprise direct torque and flux control (DTFC), direct instantaneous torque control (DITC) and direct average torque control (DATC).

The preferred method of SRM control to minimize torque ripple can be implemented by storing the torque-current-angle characteristics in a look-up table and determining the optimum current value from the rotor position and torque requirement. The DTFC method, usually called DTC, has been introduced for SRMs. It controls the torque and flux within the hysteresis bands. In the DTC method, torque is determined from the flux linkage and current. This method requires the torque values, which can be directly measured using a torque sensor or by using T-i-θ characteristics. The DATC method regulates the generated torque by controlling the phase current, turn-on and turn-off angles to maintain the desired value of torque command.

8.3.1 Design aspects

8.3.1.1 Number of poles

The "stroke angle" or "step angle" of the switched reluctance motor can be defined as follows:

$$\text{Stroke angle} = \frac{2\pi}{mN_r} \tag{8.1}$$

Where N_r = number of rotor poles, m=number of phases

The aligned position: The phase is said to be in aligned position when any pair of rotor poles is exactly aligned with stator poles of that phase. The corresponding inductance is called aligned inductance.

The unaligned position: When the inter-polar axis of the rotor is aligned with the poles of a phase, that phase is said to be in the unaligned position. The inductance at unaligned position is defined as unaligned inductance.

$$\text{Inductance ratio} = \frac{\text{Unsaturated aligned inductance}}{\text{Unsaturated unaligned inductance}} \tag{8.2}$$

The stroke angle of different stator–rotor pole configurations is shown in Table 8.2.

Lower stroke angle reduces the torque ripple, but it also reduces the inductance ratio thereby increasing the converter volt-amperes and decreasing the specific output. An SRM configuration with more rotor poles results in narrow stator poles, which leads to a fall in inductance ratio to a low value. Because of the progressive reduction of inductance ratio and energy conversion per stroke, it is not worth considering rotor pole numbers above eight for three-phase SRM.

The 12/8 configuration of three-phase SRM is a 6/4 configuration with a multiplicity of two. The 12/8 configuration has shorter end windings, which reduce the copper loss, and maintains a high inductance ratio due to the reduction in unaligned inductance. The four-pole configuration of 12/8 has the added advantage of shorter flux paths [4].

Table 8.2 Stroke Angle for Three-Phase SRM Configurations

Stator poles/Rotor poles	6/4	6/8	6/10	6/14	12/8
Stroke angle	30°	15°	12°	8.57°	15°

202 *Power electronics for electric vehicles*

8.3.1.2 Factors considered for selecting number of phases

The various factors deciding the number of phases are described in [5]:

- **Starting capability:** A PM is placed at an intermediate position of the stator pole in a single-phase motor; otherwise, it cannot start because of alignment of the stator and rotor poles.
- **Directional capability:** To run the motor in both directions, the minimum number of phases required is three.
- **Reliability:** The selection of multiphase machines is more reliable. In the event of a fault in one phase, the motor is able to run with the remaining phase for the completion of the mission, which is required for critical safety applications.
- **Cost:** The number of power electronic switches and driver circuits required increases with the selection of a higher number of phases, thereby increasing the cost of the drive.
- **Power density:** Higher power density can be achieved with the selection of a higher number of phases.
- **Efficient high-speed operation:** An SRM with a lower number of stator poles increases the efficiency at high-speed operation by reducing core losses.

Some other aspects that needs to be considered while selecting the number of phases are shown in Table 8.3.

8.3.1.3 Pole arcs

The following things are taken into consideration while selecting the pole arcs of the stator and rotor:

1. Need for self-starting.
2. To achieve proper shape of static torque with respect to rotor position.

Usually, for all practical applications, rotor poles are designed in such a way that their pole arc should be greater than or equal to the stator pole arc. The unequal pole arcs of the stator and the rotor create zero torque regions, called dead zones.

Table 8.3 Comparison of Multiphase Machines in Different Aspects

	3-phase	*4-phase*	*5-phase*
Torque dips	More	Less	Much less
Inductance ratio	High	Sufficient	Low
Multiphase excitation	Not possible	Two-phase on	Three phases on

8.3.2 Asymmetric H-bridge (AHB) converter

The accessibility for independent control of each phase with an AHB converter provides the control flexibility for SRM and eliminates the shoot-through problem. The modes of operation of the AHB converter are shown in Figure 8.7. The voltage $+V_d$ is applied across the winding if both switches S_1 (upper switch) and S_2 (lower switch) are turned on, called the magnetizing state. Turning off any one switch results in "0" voltage across the phase, called the freewheeling state. Turning off both the switches applies $-V_{dc}$ and results in demagnetization of the phase, known as the demagnetizing state.

8.3.3 Torque ripple

In recent times, SRMs have gained interest for use as variable speed drives due to their rugged construction. However, complicated control is required for servo-type applications. Moreover, torque pulsations and acoustic noise problems impact the adoption of SRMs for EV applications. In the early stage of SRM development, the drive had noisy operation with a high torque ripple. There are two ways to minimize torque ripple: improving the magnetic design and using sophisticated electronic control. In the design stage, torque ripple can be minimized with stator and rotor structure modification, and further torque ripple can be minimized by optimizing the current level, turn-on and turn-off angles, and supply voltage. The electronic

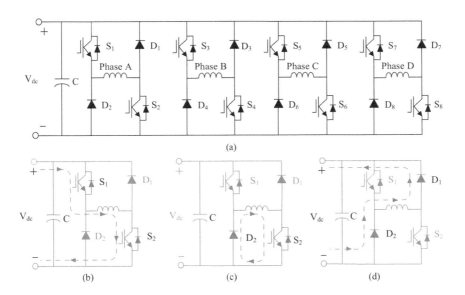

Figure 8.7 (a) AHB Converter. (b) Magnetizing mode. (c) Freewheeling mode. (d) Demagnetizing mode.

204 *Power electronics for electric vehicles*

control approach to torque ripple minimization leads to a reduction in average torque due to the lack of full utilization of motor capabilities at all power levels.

Torque ripple is inherently developed in the SRMs due to the double salient pole structure and independent excitation of phases in succession. The machine torque is defined from torque-angle-current (T-θ-i) characteristics. The amount of torque ripple is determined by the T-θ-i characteristics and magnetization pattern of individual phases [6].

8.3.3.1 Design effects on torque ripple

The torque ripple is greater in the phase overlap region during the phase commutation, where the conduction transfers from one phase to another phase. The switching frequency increases with a lower step angle due to the high number of stator–rotor pole pairs. Therefore, the design approach limits the electronic control, thereby affecting torque ripple. The amount of torque ripple of any SR motor depends on the torque dips in the T-θ-i characteristics, which are determined from the stator–rotor pole geometry, pole overlap angle, number of phases, number of poles and material properties. However, the deficiency of the incoming phase to produce sufficient torque at particular rotor positions leads to torque dips. Nonetheless, the torque dips can be eliminated by increasing the number of strokes required to complete one revolution at the penalty of saliency ratio. The ratio of maximum and minimum unsaturated inductance is referred to as the saliency ratio. The torque output decreases and volt-ampere increases with the decrease in saliency ratio, which also increases the core losses due to high switching frequency. Reducing torque dips by increasing the number of phases is the better approach in terms of smaller penalty in saliency. The average torque can be increased with a high number of phases, and the overlap region of phase in the commutation region increases, thereby reducing the torque dips. Usually, an SRM with three or lower number phases suffers from torque dips.

8.4 Torque ripple minimization

8.4.1 Torque sharing functions (TSFs)-based PWM current control

The torque control strategy is developed such that the sum of all phase torques should be constant and equal to the desired torque T_{ref}.

$$T_{\text{total}} = T_{ref} * f_T(\theta) \tag{8.3}$$

$$f_T(\theta) = \sum_{k=1}^{n} f_k(\theta) = 1 \tag{8.4}$$

where f_k is the contour function for the kth phase.

The contour function for phase 1 is

$$f_1(\theta) = \begin{cases} 0.5 - 0.5 \cos(\theta - \theta_0) & \theta_0 \le \theta < \theta_1 \\ 1 & \theta_1 \le \theta < \theta_2 \\ 0.5 + 0.5 \cos 4(\theta - \theta_2) & \theta_2 \le \theta < \theta_3 \\ 0 & \text{otherwise} \end{cases} \qquad (8.5)$$

The contour function of a phase is defined such that the function is non-zero only during the positive inductance slope region. The reference angles are selected according to the phase inductance profile of the employed motor. The conduction period of each phase is divided into three regions, which are predefined by the contour function. The phase current has conduction overlap in the first and third regions where as in second region only single phase is conducting. In this control strategy, the transition of phase current between the outgoing phase and incoming phase during commutation takes a prolonged period to achieve smooth operation. In this torque control strategy described in [7], the current in the outgoing phase is forced to follow the decaying contour instead of freewheeling, while the incoming phase current follows the increasing contour with active control.

The reference torque value is decided by the required load torque in the case of torque-controlled drives, and the current reference is taken from the readily available T-i-θ characteristics.

$$T_{total} = T_{ref}*f_1 + T_{ref}*f_2 + T_{ref}*f_3 + T_{ref}*f_4 = T_1 + T_2 + T_3 + T_4 \qquad (8.6)$$

The required current is determined using the torque equation:

$$T = \frac{1}{2}i^2\frac{dL}{d\theta} \qquad (8.7)$$

The algorithm for minimization of torque ripples with current control is as follows:

i). Obtain the torque reference value T_{ref} from the user
ii). Calculate $f(\theta_1)$ at position θ_1
iii). Calculate T_1 and T_2
iv). Obtain desired values of I_1 and I_2
v). Implement the current control to regulate the current at desired values

The desired current, which is determined from the torque equation and contour function, is

206   Power electronics for electric vehicles

$$i_d = \sqrt{\frac{2T_{ref}f(\theta)}{\dfrac{dL}{d\theta}}} \tag{8.8}$$

The condition to be satisfied to implement this current control strategy is

$$\frac{di}{d\theta} > \frac{di_d}{d\theta} \tag{8.9}$$

This condition is satisfied up to certain speeds because the rate of rising of the current versus rotor angle varies with the speed. The rate of current rise decreases with the increase in speed due to increase in back-EMF. This PWM current control is applied up to base speed because the PWM duty cycle reaches 100% at base speed. Base speed is defined as the highest speed at which maximum current can be applied to the motor at rated voltage with fixed firing angles.

8.4.2 Optimization and evaluation of TSFs for torque ripple minimization in SRM drives

Different TSFs developed for SRM are as follows which are given in [8]:

  i). Linear TSFs
  ii). Cubic TSFs
 iii). Sinusoidal TSFs
 iv). Exponential TSFs

These TSFs are developed by taking into consideration the turn-on angle, the overlap angle and the expected torque. Exponential TSFs are preferred if the objective is torque ripple minimization with maximum speed range. The preferred TSFs for two secondary objectives, such as copper loss minimization and improved speed range, are listed in Table 8.4.

Table 8.4 Selection of TSFs for Different Secondary Objectives

	Evaluation target		
	Only speed range	Only copper losses	Both speed range and copper losses
First preferred selection	Exponential TSF	Any TSF	Cubic or sinusoidal if required speed range is achieved with the rate of change of flux linkage, otherwise exponential TSF is the best selection
Second preferred selection	Cubic or sinusoidal TSF		

8.4.3 Direct torque control

DTC is a well-established torque control technique for AC machines. Over the decade it has undergone several modifications to improve its control performance. The DTC control scheme works to achieve the following objectives:

1) To maintain the constant magnitude (within hysteresis bands) of the stator flux linkage vector.
2) To maintain the electromagnetic torque in the limits of hysteresis bands by accelerating or decelerating the stator flux vector according to the requirement of torque increase or decrease. The DTC doesn't require any motor information because torque control can be achieved with the stator flux vector variation. The voltage equation representing the SRM phase is given as

$$v = R.i + \frac{d\psi(i,\theta)}{dt} \tag{8.10}$$

The stator resistance drop is negligible except at a low voltage level.
   Equation (8.10) can be rewritten as

$$\frac{\overline{d\psi}}{dt} \approx \vec{v} \tag{8.11}$$

This equation shows that the flux varies in direct proportion to the voltage vector, and the magnitude of the flux vector depends on the voltage vector magnitude and the time interval for which that voltage vector is in the active state [9].

8.4.3.1 Voltage vector selection

The stator flux and torque can be controlled directly by selecting the proper voltage vector. The selection of the voltage vector is as follows:

- A voltage vector that makes an angle that does not exceed $\pm 90°$ with the stator flux vector is selected to increase the flux.
- A voltage vector that makes an angle that does exceed $\pm 90°$ with the stator flux vector is selected to decrease the flux.
- The torque can be increased and decreased by accelerating and decelerating the stator flux vector, respectively.

Four different possibilities exist for the outputs of torque and flux hysteresis controllers. The symbol T↑ represents that torque error is positive, i.e., reference torque is more than actual torque, which needs an increase in torque, and T↓ represents the need for decrease in torque. Similarly, Ψ↑ and Ψ↓ represent the need for increase and decrease of stator flux, respectively.

8.4.3.2 DTC of four-phase SRM

The block diagram of DTC for a four-phase SRM is shown in Figure 8.8. The mathematical model of SRM is developed using flux linkage and torque characteristics in the look-up table approach. The flux linkage and torque characteristics are obtained from experiments. The characteristics data obtained from the experimental test is more suitable for modelling SRM than the finite element analysis (FEA) approach due to the exclusion of secondary effects in FEA. The selection of voltage vectors for the four-phase SRM is shown in Table 8.5. The sector diagram for four-phase is shown in Figure 8.9.

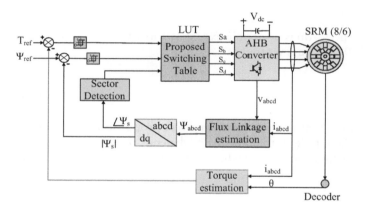

Figure 8.8 Block diagram of DTC of four-phase SRM.

Table 8.5 Voltage Vector Selection for Conventional DTC

Sector	T↑Ψ↑	T↑Ψ↓	T↓Ψ↑	T↓Ψ↓
N_1	V_2	V_4	V_8	V_6
N_2	V_3	V_5	V_1	V_7
N_3	V_4	V_6	V_2	V_8
N_4	V_5	V_7	V_3	V_1
N_5	V_6	V_8	V_4	V_2
N_6	V_7	V_1	V_5	V_3
N_7	V_8	V_2	V_6	V_4
N_8	V_1	V_3	V_7	V_5

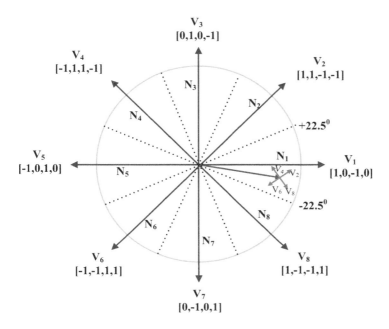

Figure 8.9 Voltage vector selection for conventional DTC.

The performance of the SRM drive under the DTC method is verified with MATLAB simulations conducted on 4 kW, four-phase 8/6 SRM. The simulations are performed at 500 rpm, 5 Nm reference torque, and flux linkage reference of 0.6 Weber-sec. The simulation waveforms of electromagnetic torque developed T_{em}, phase currents i_{abcd}, phase torques T_{abcd} and sectors for conventional DTC are shown in Figure 8.10. From Figure 8.14, it is shown that negative torque is generated in each phase because of current drawn by phase in the negative slope phase inductance region. In addition to this, conventional DTC maintains the current in the dead zone (zero torque region) as shown in Figure 8.12, which leads to a reduction in the T/A ratio. The T/A ratio is one of the performance indexes of the SRM, and it indirectly indicates the efficiency [10]. A considerable amount of torque ripple exists in the conventional DTC due to improper voltage vector selection to maintain the desired values of torque and flux.

The voltage across and current through phase A winding, along with the gate signals, are presented in Figure 8.11. From Figure 8.12, it can be observed that motor phase winding carries the current during the negative torque region as well as the zero torque region. The existence of current in the negative and zero torque regions decreases the T/A ratio and also results in high torque pulsations.

210 Power electronics for electric vehicles

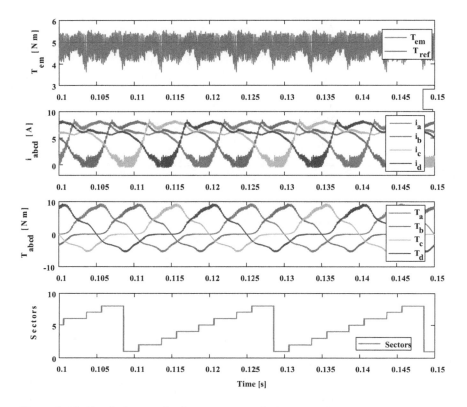

Figure 8.10 Simulation results for conventional DTC at 500 rpm.

8.4.4 DTC – 8-vector 16-sector partition method for torque ripple minimization

The optimized performance and efficient control of SRM can be achieved through torque ripple minimization and by enhancing the T/A ratio. The T/A ratio can be improved by limiting the current drawn from the source to develop the required amount of torque by eliminating the current in the zero torque regions. This can be achieved with voltage vector selection optimization and sector reorganization. For the precise selection of voltage vectors, 8 sectors of the DTC algorithm are divided into 16 sectors [11]. Furthermore, this approach facilitates the selection of vectors that are at right angles to the flux linkage vector to increase/decrease the torque with a faster dynamic response. This can be achieved in the 16-sector partition method due to the change in voltage vector section for every 22.5° of rotor position, which happens for every 45° in conventional DTC.

Selection of electric drive for EVs 211

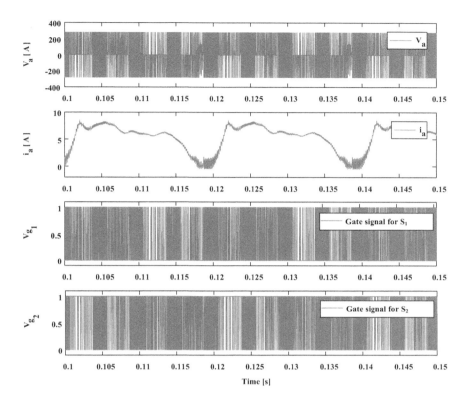

Figure 8.11 Simulation results for conventional DTC corresponding to phase A.

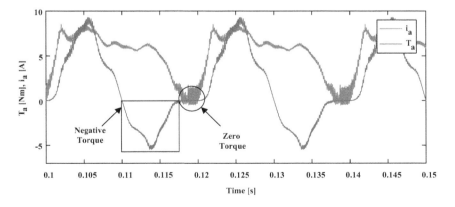

Figure 8.12 Interpretation of zero and negative torque regions in conventional DTC.

212 Power electronics for electric vehicles

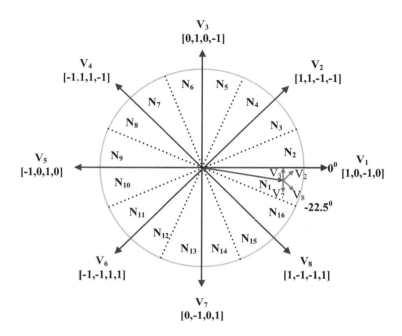

Figure 8.13 Definition of voltage vectors and partition of sectors for proposed DTC.

The vector diagram of the proposed DTC method with 16-sector partition is depicted in Figure 8.13, and the voltage vector selection is given in Table 8.6. The simulation results of the proposed 8-vector 16-sector partition method are shown in Figure 8.14.

The voltage and current corresponding to phase A winding along with the gate signals are shown in Figure 8.15. From Figure 8.16, it can be observed that the existence of current in the zero torque region is eliminated, and therefore, the proposed method enhances the T/A ratio.

The dynamic response comparison of the conventional 8-sector method and the proposed 16-sector partition method is presented in Figure 8.17. The proposed DTC has a fast dynamic response compared with conventional DTC.

The comparison of torque ripple and T/A of 8- sector and 16-sector partition methods are shown in Table 8.7. The 16-sector method has a better performance than the 8-sector method due to optimized voltage vector selection.

Table 8.6 Voltage Vector Selection for Proposed 8-Vector 16-Sector DTC Method

Sector	$T\uparrow\Psi\uparrow$	$T\uparrow\Psi\downarrow$	$T\downarrow\Psi\uparrow$	$T\downarrow\Psi\downarrow$
N_1	V_2	V_3	V_8	V_7
N_2	V_3	V_4	V_1	V_8
N_3	V_3	V_4	V_1	V_8
N_4	V_4	V_5	V_2	V_1
N_5	V_4	V_5	V_2	V_1
N_6	V_5	V_6	V_3	V_2
N_7	V_5	V_6	V_3	V_2
N_8	V_6	V_7	V_4	V_3
N_9	V_6	V_7	V_4	V_3
N_{10}	V_7	V_8	V_5	V_4
N_{11}	V_7	V_8	V_5	V_4
N_{12}	V_8	V_1	V_6	V_5
N_{13}	V_8	V_1	V_6	V_5
N_{14}	V_1	V_2	V_7	V_6
N_{15}	V_1	V_2	V_7	V_6
N_{16}	V_2	V_3	V_8	V_7

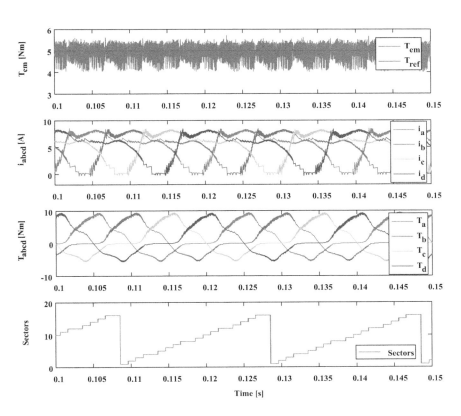

Figure 8.14 Simulation results for proposed DTC at 500 rpm.

214 *Power electronics for electric vehicles*

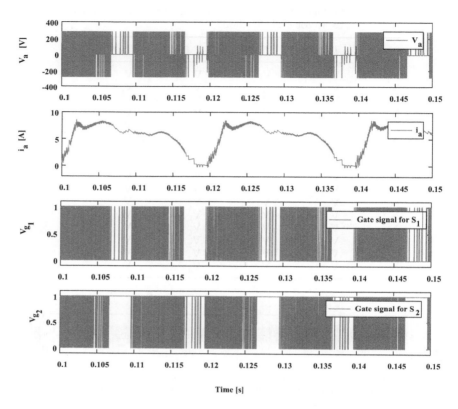

Figure 8.15 Simulation results for proposed DTC corresponding to phase A.

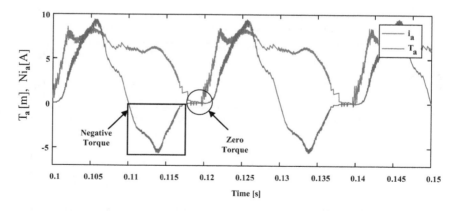

Figure 8.16 Interpretation of zero and negative torque regions in proposed DTC.

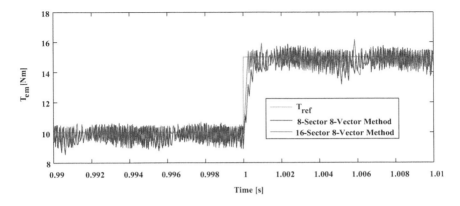

Figure 8.17 Dynamic response comparison of conventional DTC and proposed DTC.

Table 8.7 Comparison of Conventional and Proposed 8-Vector 16-Sector DTC Methods

Speed (rpm)	Torque ripple (%)		T/A (Nm/A)	
	Conventional DTC	Proposed DTC	Conventional DTC	Proposed DTC
100	26.8	25.1	0.8107	1.2041
250	31.1	28.1	0.8289	1.1804
500	38.4	33.1	0.8321	1.1071

References

1. B. Bilgin, J. W. Jiang and A. Emadi (Eds.), *Switched Reluctance Motor Drives: Fundamentals to Applications* (1st ed.). CRC Press, 2018.
2. B. Bilgin and A. Emadi, "Electric Motors in Electrified Transportation: A Step Toward Achieving a Sustainable and Highly Efficient Transportation System," *IEEE Power Electronics Magazine*, vol. 1, no. 2, pp. 10–17, Jun. 2014.
3. I. Husain, *Electric and Hybrid Vehicles: Design Fundamentals* (3rd ed.). CRC Press, 2021.
4. T. J. E. Miller, *Switched Reluctance Motors and Their Control (Monographs in Electrical and Electronic Engineering)*. Clarendon Press, 1993.
5. R. Krishnan, *Switched Reluctance Motor Drives: Modeling, Simulation, Analysis, Design, and Applications*. CRC Press, 2001.
6. I. Husain, "Minimization of Torque Ripple in SRM Drives," *IEEE Transactions on Industrial Electronics*, vol. 49, no. 1, pp. 28–39, Aug. 2002.

216 *Power electronics for electric vehicles*

7. I. Husain and M. Ehsani, "Torque Ripple Minimization in Switched Reluctance Motor Driven by PWM Current Control," *IEEE Transactions on Power Electronics*, vol. 11, no. 1, pp. 83–88, Jan. 1996.
8. X. D. Xue, K. W. E. Cheng and S. L. Ho, "Optimization and Evaluation of Torque-Sharing Functions for Torque Ripple Minimization in Switched Reluctance Motor Drives," *IEEE Transactions on Power Electronics*, vol. 24, no. 9, pp. 2076–2090, Sept. 2009.
9. A. D. Cheok and Y. Fukuda, "A New Torque and Flux Control Method for Switched Reluctance Motor Drives," *IEEE Transactions on Power Electronics*, vol. 17, no. 4, pp. 543–557, July 2002.
10. N. Yan, X. Cao and Z. Deng, "Direct Torque Control for Switched Reluctance Motor to Obtain High Torque–Ampere Ratio," *IEEE Transactions on Industry Applications*, vol. 66, no. 7, pp. 5144–5152, July 2019.
11. P. K. Reddy, D. Ronanki and P. Parthiban, "Direct Torque and Flux Control of Switched Reluctance Motor with Enhanced Torque Per Ampere Ratio and Torque Ripple Reduction," *Electronic Letters*, vol. 55, no. 8, pp. 477–478, 2019.

9 Voltage lift quasi Z-source inverter topologies for electric vehicles

Josephine R L, Gollapinni Vaishnavi, Akshatha Patil, M Sai Harsha Naidu, and Rachaputi Bhanu Prakash

9.1 Introduction

Nowadays, electric vehicles (EVs) are gaining a lot of popularity because of their increased performance and being environmentally friendly. Energy storage systems (ESSs) are one of the important components used in EVs for power supply. The types of ESSs are battery storage systems (BSSs), supercapacitors and fuel cells. BSSs are smaller in size and have high energy density, thus operating efficiently with an increased life span. Supercapacitors have fast charging and discharging capacity. However, their energy density is low. Fuel cells come with high energy density, but they have drawbacks of having high cost and non-supportive infrastructure.

Proper functioning of the EV is possible with the right interface of the ESS to the motors used in EVs. However, the output of the ESS is unregulated, and also, there would be a voltage drops [1, 2]. As a result, the direct current (DC) output voltage will have low magnitude, and this necessitates increasing the voltage. To amplify the value of DC voltage, DC–DC regulators are used, which will contribute to the increased weight and space. In this chapter, the emphasis is on designing an inverter that has an implicit ability to provide voltage gain, thus eliminating the need for a DC–DC converter and helping to reduce the size, weight and volume for EV application.

The voltage at the output side of standard VSI [3] is adjusted using the pulse width modulation (PWM) approach. However, both switches on the same leg of the inverter topology cannot be switched on at the same time. If they are, it creates a shoot-through problem. To overcome the difficulty involved in traditional VSIs, inverters of single-stage type are proposed with both buck and boost capability [4, 5]. A new Z-source inverter (ZSI) (Figure 9.1) was first proposed in Peng [6], where it uses a unique X-shaped passive network, which is made up of two inductances and two capacitances. This network is instrumental in giving the boost factor. This uses shoot-through zero state and eliminates dead time, minimizing the waveform distortion. In Peng [6] and Loh et al. [7], various control strategies are

DOI: 10.1201/9781003248484-9

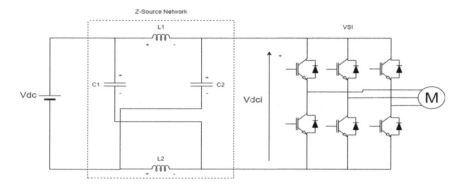

Figure 9.1 Z-source inverter.

proposed based on conventional PWM methods that are compatible with ZSI.

Practically, more emphasis is given to three-phase motors because of their good speed range. Single-phase inductor motors have many attractive features, but they have some drawbacks, such as difficulty in starting and short speed range. To overcome these difficulties, Vodovozov et al. [8] have proposed methods to improve speed range to meet the requirements of EVs. The single-phase motors are well suited for two-wheelers and electric bicycles.

Fuel cell electric vehicles (FCEVs) have received a lot of interest recently [9–15] because they have a few important benefits over battery electric vehicles (BEVs). The battery charging period is long; however, with an FCEV, the gasoline tank can be filled in a fraction of the time it takes in a standard internal combustion engine (ICE) car. There are no emissions from an FCEV. However, due to the fuel cells' inability to support regenerative braking, energy is lost when braking. As a result, there is a need for a separate energy storage system [22]. A battery or a supercapacitor can be used. Furthermore, the voltage that is available from the fuel cell arrangement is low (about 150–300 V) and unregulated, which is insufficient to power the electric vehicle's propelling machine.

Motors for EVs are typically rated at 400 V. However, in present times, attention has switched to 800 V machines [16]. In comparison with 400 V machines, 800 V machines require less current for the same power because the latter have a greater voltage. As a result, the copper wire inside the motor is smaller, and hence, the motor's mass and volume can be reduced. In addition, because the needed current is lower, the variable losses (I^2R) are lower. As a result, as compared with a 400 V machine, the efficiency is likewise greater. These are the main benefits of using an 800 V motor.

Thus, if the input energy supply is a fuel cell setup, and a machine rated at 800 V is employed, a high voltage gain is expected [17, 18]. ZSI Figure 9.1

Voltage lift quasi Z-source inverter 219

and QZSI have a tiny boost factor. Furthermore, the voltage boost range is limited by parasitic elements and may not be as intended. High gain sepic-based ZSI and voltage lift Quasi ZSI are offered as a result.

9.2 High gain sepic-based Z-source inverter

As said, the inverter is formed by combining a modified sepic converter, which has a switched inductor in it, with the ZSI. The total inverter circuit (Figure 9.2) functions like a single-stage inverter with an implicit boost capacity.

Switch S1 is responsible for boosting the voltage, whereas switch S2 is responsible for producing a sinusoidal output voltage. Hence, at any time, only two switches must be turned on. Only when one of the switches, S1 or S2, is switched on is switch S3 activated.

9.2.1 Mode 1

The following are the equations corresponding to mode 1:

$$v_{Lf_1} = v_{in}, v_{Lf_2} = v_{in}, v_{L_1} = -\left(v_{c_0} + v_{c_2} + v_{C_3}\right), v_{L_2} = -\left(v_{c_0} + v_{c_1} + v_{C_3}\right), v_{L_3} = -\left(v_{C_3}\right)$$

(9.1)

$$\dot{i}_{C_1} = i_{L_2}, \dot{i}_{C_2} = i_{L_1}, \dot{i}_{C_3} = i_{L1} + i_{L_2} + i_{L_3}, \dot{i}_{C_0} = i_{L1} + i_{L_2} - i_0$$

(9.2)

9.2.2 Mode 2

The following are the equations corresponding to mode 2:

$$v_{Lf_1} = v_{in}, v_{Lf_2} = v_{in}, v_{L_1} = v_{C_2}, v_{L_2} = v_{C_1}, v_{L_3} = -v_{C_3}$$

(9.3)

$$\dot{i}_{C_1} = -i_{L_1}, \dot{i}_{C_2} = -i_{L_2}, \dot{i}_{C_0} = -i_0$$

(9.4)

Figure 9.2 depicts the circuit diagram of the high gain sepic-based ZSI, Figure 9.3 depicts mode 1, Figure 9.4 depicts mode 2 and Figure 9.5 depicts mode 3.

During mode 1, switches S1 and S3 are closed, and switch S2 is opened. Magnetizing inductances L_{f1} and L_{f2} are charged from source. Capacitors C_1, C_2 and C_3 will be charged. The switching devices S1 and S2 are closed during mode 2, but switch S3 is opened. The inductors L_{f1} and L_{f2} are magnetized by source voltage. The capacitors C_1, C_2 and C_3 release the stored energy. During mode 3, the switches S2 and S3 are closed, whereas the switch S1 is open. In this last mode, the magnetizing inductances are demagnetized, and the

220 *Power electronics for electric vehicles*

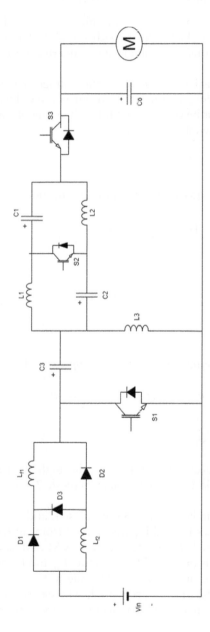

Figure 9.2 Modified sepic converter with switched inductor and Z-source inverter.

Voltage lift quasi Z-source inverter 221

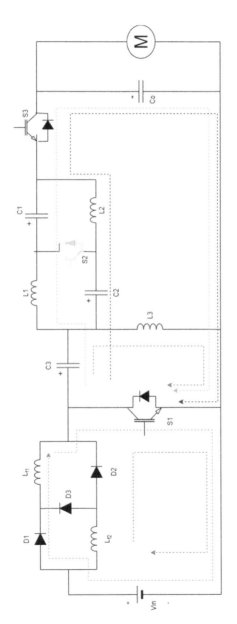

Figure 9.3 Mode 1.

222 Power electronics for electric vehicles

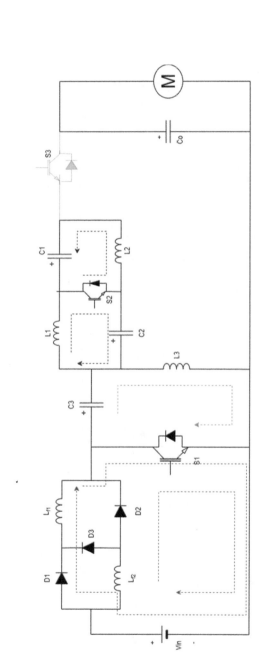

Figure 9.4 Mode 2.

Voltage lift quasi Z-source inverter 223

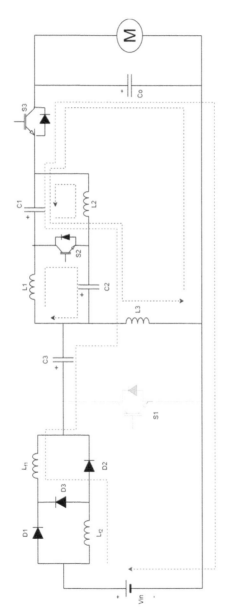

Figure 9.5 Mode 3.

224 *Power electronics for electric vehicles*

capacitors are charged. The equations governing these are written below each mode using Kirchhoff voltage law (KVL) and Kirchhoff current law (KCL).

9.2.3 Mode 3

The following are the equations corresponding to mode 3:

$$v_{Lf_1} = 0.5\left(v_{in} - \left(v_{C_0} + v_{C_1} + v_{C_2} + v_{C_3}\right)\right), v_{L_1} = v_{C_1}, v_{L_2} = v_{C_2}, v_{L_3} = v_{C_0} + v_{C_1} + v_{C_2}$$

$$(9.5)$$

$$i_{C_1} = i_{Lf_1} + i_{Lf_2} - i_{L_1} - i_{L_3}, i_{C_2} = i_{Lf_1} + i_{Lf_2} - i_{L_2} - i_{L_3} \tag{9.6}$$

$$i_{C_3} = i_{Lf_1} + i_{Lf_2}, i_{C_0} = i_{Lf_1} + i_{Lf_2} - i_{L_3} - i_0 \tag{9.7}$$

where v_{C_1}, v_{C_2}, v_{C_3} and v_{c_0} denote the voltage of the capacitors C_1, C_2, C_3 and C_0, respectively. The currents through inductors L_1, L_2, L_3 and L_f are $i_{L_1}, i_{L_2}, i_{L_3}, i_{Lf}$, respectively.

The voltage and current gains are calculated using the volt-second and ampsecond balance principles. The following are the corresponding equations.

$$\frac{V_{C_1}}{V_{in}} = \frac{Vc_2}{V_{in}} = \frac{\left(1 + 3D_1\right)\left(1 - D_2\right)}{\left(1 - D_1\right)} \tag{9.8}$$

$$\frac{V_{C_3}}{V_{in}} = \left(1 + 3D_1\right) \tag{9.9}$$

$$\frac{V_0}{V_{in}} = \frac{\left(1 + 3D_1\right)\left(D_1 + 2D_2 - 2\right)}{\left(1 - D_1\right)} \tag{9.10}$$

$$i_{L_1} = i_{L_2} = i_0 \tag{9.11}$$

$$i_{L_3} = -i_0 \tag{9.12}$$

$$i_{Lf} = \frac{2 - D_1 - 2D_2}{D_1 - 1} i_0 \tag{9.10}$$

9.3 Design of inductors

Based on the equations obtained in mode 1, where the magnetizing inductance is getting magnetized by the source at the input side and considering ripple content in the inductor, the magnetizing inductances L_{f1} and L_{f2} are calculated as follows.

$$L_{f1} = L_{f2} = \frac{V_{in}D_1T_s}{\Delta i_{Lf}} \tag{9.14}$$

where D_1 is the duty factor for switch S1, T_s denotes switching time duration. The ripple current in inductors is denoted by Δi_{Lf}.

Similarly, the design of inductors L_1 and L_2 is related to the duty cycle D_2. Considering the ripple in those inductors, the inductors are designed as follows, and for simplicity, the values of both the inductors are taken to be same.

$$L_1 = L_2 = \frac{V_{in}\left(1 + 3D_1\right)\left(1 - D_2\right)D_2T_s}{\left(1 - D_1\right)\Delta iL_1} \tag{9.15}$$

By combining Eqs (9.6) and (9.7), L_3 is designed as follows,

$$L_3 = \frac{V_{in}\left(1 + 3D_1\right)D_1T_s}{\Delta i_{L_3}} \tag{9.16}$$

9.4 Design of capacitors

The value of C_1, the current in the inductor L_1, the switching time period T_s, and the duty cycle of switch S_2 all influence the voltage ripple in the capacitor C_1. Using $i_C = C \cdot \dfrac{dv}{dt}$ and by applying the same principle as that of inductors, the capacitance is calculated as

$$C_1 = C_2 = \frac{i_{L_1}\left(1 - D_2\right)T_s}{\Delta v_{C_2}} \tag{9.17}$$

where Δv_{C_2} is the ripple in the capacitor C_2. Similarly, the capacitor C_3 is designed by considering ripple voltage Δv_{C_3} in C_3. The capacitor design C_3 is also dependent on the current of inductor L_3. The equation is given as

$$C_3 = \frac{i_{L_3}\left(1 - D_2\right)T_s}{\Delta v_{C_3}} \tag{9.18}$$

Finally, the output capacitor C_0 is calculated as

$$C_0 = \frac{2I_0D_3T_s}{\Delta v_{C_0}} \tag{9.19}$$

226   *Power electronics for electric vehicles*

where D_3 denotes duty ratio of switch S3 and v_{C_0} denotes the ripple content in the output capacitor.

9.5 Control scheme

A sepic converter and a semi-ZSI are coupled in this circuit. To create the new architecture, a capacitor swaps the input DC source voltage in the basic semi ZSI.

$$v_0 = \frac{2D-1}{1-D} v_C$$

It's also possible to change this equation and write it as

$$v_0 = (2D-1)(2v_C + v_0)$$

$2v_c + v_0$ is held constant in this calculation, and output voltage and duty factor D have a linear connection. The switching drive sequence must satisfy the following restrictions to keep $2v_c + v_0$ constant: when S1 is turned on, one of the switches S2 or S3 must be turned off. Both S2 and S3 must be switched on while S1 is inactive. When S2 is activated, at least one of the S2 or S3 switches should be turned off, and when S2 is shut off, both S2 and S3 should be turned on. This is utilized in the switching scheme.

The sepic-based ZSI's AC output voltage is defined as

$$v_0 = V_0 \sin \omega t \tag{9.20}$$

which can be further written as $v_0 = A V_{in} \sin wt$, where A denotes peak voltage gain and V_0 denotes the peak output voltage.

For boosting the value of the voltage, D_1 is maintained constant, and the boost factor k is

$$k = \frac{D_1(1+D_1)}{(1-D_1)} \tag{9.21}$$

To obtain the sinusoidal output voltage, duty ratio D_2 is varied as a function of sine, and it is derived considering the voltage gain equation of the inverter:

$$D_2 = 1 - 0.5D_1 + 0.5\left(\frac{1-D_1}{1+3D_1}\right) A \sin wt \tag{9.22}$$

D3 denotes the duty ratio of switch S3, which will be activated only if one of the other switches, S1 or S2, is activated. This logic is implemented by using XOR gate. The duty ratio D_3 for S3 is given here.

Voltage lift quasi Z-source inverter 227

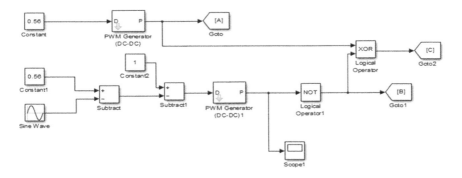

Figure 9.6 Control scheme simulated on MATLAB Simulink.

$$D_3 = 2 - D_1 - D_2 \quad (9.23)$$

For boosting the voltage, the boost factor k is kept constant, and the result is compared with the carrier signal to generate pulses for switch S1. Switch S2 is responsible for creating sinusoidal output voltage. To acquire switching pulses for S2, the signal denoted by Eq. (9.22) is compared with the carrier signal. Using XOR logic, switching pulses are generated for switch S3, as this switch is to be switched on only when one of the switches S1 or S2 is in conduction. In this way, by following a three-switch three-state control scheme, we can successfully obtain sinusoidal voltage at the output. The control scheme simulated on MATLAB is shown in Figure 9.6.

9.6 Experimental analysis

The proposed inverter model was tested using MATLAB Simulink for a 120 V rms output at 50 Hz. The switching frequency is considered to be 30k Hz. Inductor ripple current can be estimated to be between 20% and 40% of the total current. As a result, a value of 20% of the current is considered. The simulation was run at various switching frequencies, with the ripple content being determined to be lower at 30 KHz. The equations developed previously are used to compute capacitances and inductances. Input voltage is considered as 75 V DC. The ripple for inductor current is considered as 20%, and the ripple for capacitor voltage is considered to be 7%. The motor is rated at 0.5 Hp. To make the calculations simpler, the two inductances and the two capacitance values are considered to be the same. By trial and error, the value of the boost factor k is set to be 2, and the value of inverter gain factor comes out to be 2.26. The switches used are insulated gate bipolar transistor (IGBT.) Table 9.1 shows the simulation parameter values.

The switch stress is stated in terms of voltage and current as follows, using Eqs (9.1), (9.2) and (9.7)–(9.12):

228　*Power electronics for electric vehicles*

Table 9.1 Simulation Parameter Values

Parameter	Value
Input voltage, V_{in}	75 V
Output voltage, V_0	120 V
Switching frequency, f_s	30 kHz
Output voltage gain, A	2.26
Maximum boost factor, k	2
Inductance Lf1 = Lf2	1.54 mH
Impedance network inductances L1 and L2	1.29 mH
Impedance network capacitances C1 and C2	5.60 μF
Capacitance C3	8 μF
Inductance L3	2.61 mH
Output capacitance Co	23.5 μF

$$V_S = (1 + K)V_{in} \text{ and } I_s = (1 + A)I_0$$

The boost ratio is K, while the voltage gain ratio is A. Output AC current is I0. Switches should be chosen based on the switch voltage and current stress.

9.7　Circuit analysis of VL-QZSI

VL-QZSI, with passive components connected between the DC supply and a typical VSI network, is shown in Figure 9.9. The passive components are three inductances, three capacitances and three diodes.

In the stated VL-QZSI circuitry, there are two operating modes: shoot-through mode (Figure 9.12) and non-shoot-through mode (Figure 9.13).

9.7.1　Shoot-through mode

Voltage lift cell shown in Figure 9.11 is incorporated in the quasi Z-source inverter Figure 9.10. The two unused states in a VSI, during which only upper switches of each pole or lower switches of each pole are turned ON, will be turned into the shoot-through state, in which every switching device is activated. Figure 9.12 depicts the circuit that corresponds to this working mode.

In this mode, diode D1 doesn't operate, while diodes D2, D3 and D4 operate. Through capacitor C2 as well as the DC source, L1 stores energy. Inductors L2 and L3 and capacitor C3 are receiving energy from capacitor C1.

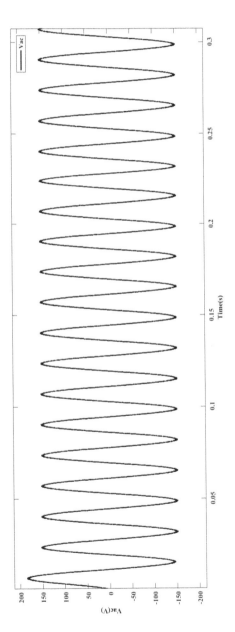

Figure 9.7 Output voltage waveform.

230 Power electronics for electric vehicles

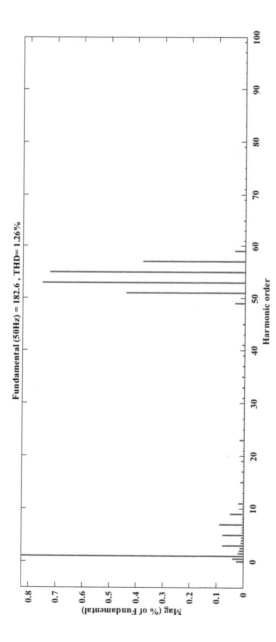

Figure 9.8 THD of output AC voltage.

Voltage lift quasi Z-source inverter 231

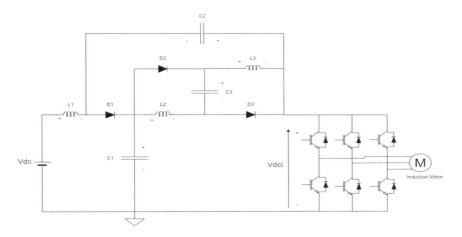

Figure 9.9 VL-QZSI.

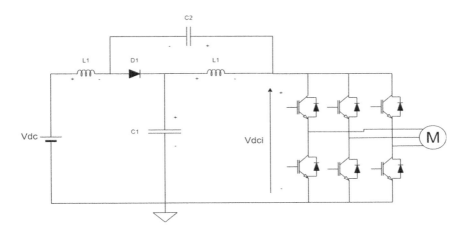

Figure 9.10 QZSI.

The equations that apply to this mode are as below:

$$v_{L1} = V_{dc} + v_{c2}, v_{L2} = v_{c1} = v_{L3} = v_{c3}, V_{dci} = 0 \tag{9.24}$$

$$i_{c1} = i_{L2} + i_{L3} + i_{c3}, i_{c2} = i_{L1}, i_{c3} = i_{c1} - i_{L2} - i_{L3} \tag{9.25}$$

where V_{dc} denotes DC voltage, which is not regulated, and the maximum value of boosted DC voltage, V_{dci}, serves as the input for the traditional VSI section of the circuitry.

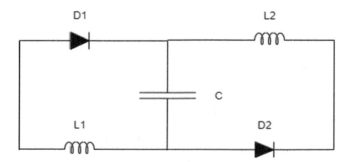

Figure 9.11 Voltage lift cell.

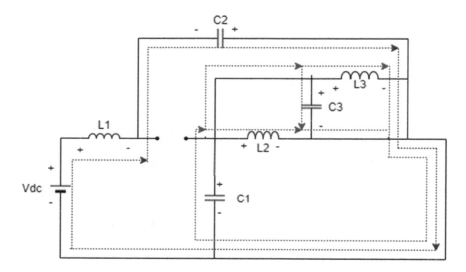

Figure 9.12 Shoot-through mode of operation.

9.7.2 Non-shoot-through mode

In this, the inverter circuit is like a typical VSI. D1 operates, and D2 and D3 don't operate. Voltage across the inductances L_1, L_2 and L_3 and capacitance C_3 are added with V_{dc}. As a result, V_{dci} (boosted voltage) acts as the input to the inverter section. The capacitors C_1 and C_3 are now storing energy. Figure 9.10 depicts the corresponding circuit for non-shoot-through mode.

Voltage lift quasi Z-source inverter 233

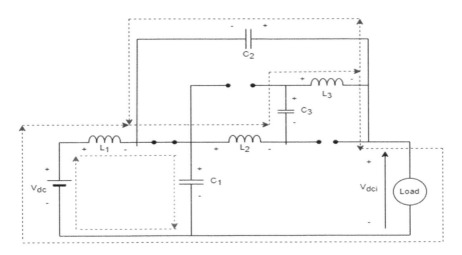

Figure 9.13 Non-shoot-through mode.

The equations that correspond to this mode are as follows:

$$v_{L1} = V_{dc} - v_{c1}, v_{c2} = v_{c3} - v_{L2} - v_{L3}, V_{dci} = v_{c1} + v_{c2} \tag{9.26}$$

$$i_{c1} = i_{L1} - i_{L2} + i_{c2}, i_{c2} = i_{L2} - I_{load}, i_{c3} = i_{L2} \tag{9.27}$$

Inductances L_2 and L_3 operate in series, and a presumption that L_2 and L_3 are equal is made. Hence, $v_{L2} = v_{L3}$. So, from Eq. (9.26),

$$v_{L2} = \frac{v_{c3} - v_{c2}}{2} \tag{9.28}$$

Through the volt–sec balance concept, the average voltages across capacitances C_1, C_2 and C_3 are

$$V_{c1} = V_{c3} = \frac{1-d}{1-3d} * V_{dc}, V_{c2} = \frac{1+d}{1-3d} * V_{dc} \tag{9.29}$$

The peak value of the boosted DC voltage (V_{dci}) can be derived from Eq. (9.26) as shown here:

$$V_{dci} = \frac{2}{1-3d} * V_{dc} = B * V_{dc} \tag{9.30}$$

where B is the boost factor, while d is the shoot-through duty ratio.

234 *Power electronics for electric vehicles*

The amplitude of the phase voltage $(\hat{v}_{ac})$ that would be output from the DC–AC converter is given here:

$$\hat{v}_{ac} = m * \frac{V_{dci}}{2} = m * B * \frac{V_{dc}}{2} = G * \frac{V_{dc}}{2} \tag{9.31}$$

9.8 Design of passive components (L and C)

The relationship between average inductor current and load current is required to determine inductance levels. The following expressions are generated by applying the amp–sec balance concept to the current expressions in Eqs (9.25) and (9.27):

$$I_{L1} = I_{load}\left[\frac{1 - 5d + 4d^2}{1 - 2d - 2d^2}\right], I_{L2} = I_{load}\left[\frac{1 - 3d + 2d^2}{1 - 2d - 2d^2}\right] \tag{9.32}$$

The ripple component in the inductor current is taken as $y\%$ of the average inductor current. Considering $v_L = L \cdot \frac{di}{dt}$ and Eq. (9.29),

$$L_1 = \frac{(V_{dc} + V_{c2}) * d}{y\% * I_{L1} * f_{sw}}, L_2 = \frac{V_{c1} * d}{y\% * I_{L2} * f_{sw}} = L_3 \tag{9.33}$$

The ripple component in the capacitor voltage is taken into account as $y\%$ of the average capacitor voltage when determining capacitance values, and considering $i_C = C \cdot \frac{dv}{dt}$ and Eqs (9.25) and (9.27),

$$C_1 = \frac{I_{c1} * d}{y\% * V_{c1} * f_{sw}}, C_2 = \frac{I_{c2} * d}{y\% * V_{c2} * f_{sw}}, C_3 = \frac{I_{c3} * d}{y\% * V_{c3} * f_{sw}} \tag{9.34}$$

where f_{sw} is the switching frequency.

9.9 Switching scheme

The sinusoidal pulse width modulation (SPWM) technique is commonly employed for VSI switching. When the triangular carrier signal is greater than all sine reference signals, all the bottom switching devices in each pole of the VSI will be turned ON for a certain period of time, and while the triangular carrier signal is lower than all sine reference signals, all the top switching devices in each pole of the VSI will be turned ON for a certain period of time. The ZSI converts these inactive states to shoot-through states by sending gating pulses to top switching devices when all bottom switching devices are ON, and vice versa. Modified PWM approaches are used to generate gating pulses, and these techniques use two additional modulating signals [19–21].

In this work, maximum constant boost PWM (MCB-PWM) strategy (Figure 9.14) is taken into account [10, 14]. V_p and V_n, the two additional modulating signals, are also sine signals. The equations that govern the signals are as follows:

$$V_p = \sqrt{3} * m + m * \sin\left(\theta - \frac{2\pi}{3}\right) \quad \text{for } 0 < \theta < \frac{2\pi}{3}$$

$$V_p = \sin(\theta) * m \quad \text{for } \frac{\pi}{3} < \theta < \frac{2\pi}{3} \quad (9.35)$$

The following are equations that govern V_n:

$$V_n = \sin\left(\theta - \frac{2\pi}{3}\right) * m \quad \text{for } 0 < \theta < \frac{2\pi}{3}$$

$$V_n = \sin(\theta) * m - \sqrt{3} * m \quad \text{for } \frac{\pi}{3} < \theta < \frac{2\pi}{3} \quad (9.36)$$

The time period of signals is one-third that of the real sinusoidal modulating signals, as seen in the definitions earlier. The gap between two additional reference curves is (Vp, Vn) constant, as is shown. This leads to a consistent shoot-through duty ratio (d) for any given modulation index (m), as opposed to the max boost PWM approach, which causes a ripple component

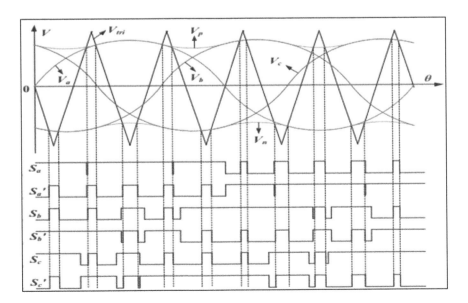

Figure 9.14 Schematic of MCB-PWM.

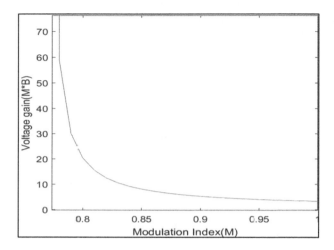

Figure 9.15 G versus m.

in inductor current and capacitor voltage. The correlation between d and m is given by $= 1 - \frac{\sqrt{3}m}{2}$.

Variability of the voltage gain (G) from the input DC side to the output AC side is shown in Figure 9.15. When compared with other similar PWM approaches in [6, 19], the modulation index is higher for any desired gain value. The benefit is that for the same value of m, d is decreased, leading to reduced switch conduction losses. The implementation of the switching method is depicted in a simple block diagram as in Figure 9.16.

9.10 Experimental analysis

Table 9.2 presents the input and intended output conditions as well as the values of several impedance network elements.

The switching frequency is set at 10 kHz and the output power as 100 kW, Using values in Table 9.2, the boosted DC voltage (Figure 9.17a and b) and the AC output (Figure 9.17c) were obtained.

The switches used in the inverter topology are IGBTs. The modulation index corresponding to the considered voltages is $M = 0.93532$. The value of boosted DC voltage is 1115.7 V, and the AC peak phase voltage is 648.1 V. The ripple component of 10% of the average inductor current is taken into account while designing inductances. Similarly, the 1% ripple component in the average capacitor voltage is taken into account while designing capacitances. After simulating the model voltage waveform at the output is as

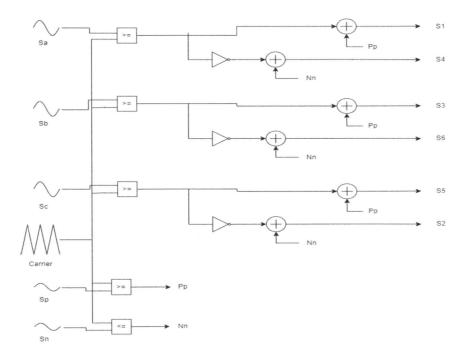

Figure 9.16 Schematic of switching scheme implementation.

Table 9.2 Parameters of the Impedance Network

Parameter	Value
Input Voltage	300 V
AC Output Voltage (V_{rms})	800 V
Inductance (L_1)	7.544 mH
Inductance (L_2)	4.08 mH
Inductance (L_3)	4.08 mH
Capacitance (C_1)	267.1 µF
Capacitance (C_2)	65.14 µF
Capacitance (C_3)	1053.7 µF

shown in Figure 9.7 and its corresponding total harmonic distortion is as shown in Figure 9.8.

Figure 9.17b shows the boosted DC voltage (V_{dci}). During the transient period, this voltage has a significant peak. This is because of the L and C component values. The L and C components are designed based on the considered ratings. These values are causing the system to overshoot. As a

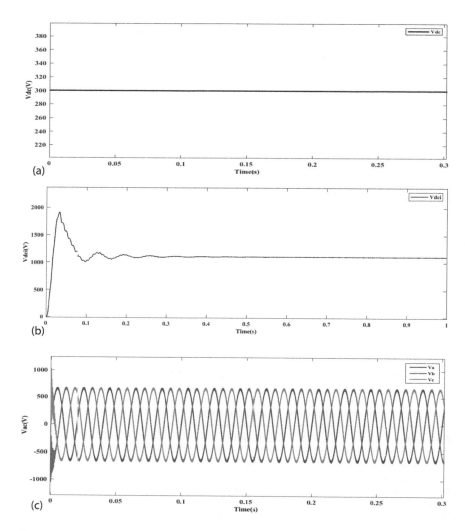

Figure 9.17 (a) DC input voltage. (b) Boosted DC voltage (V_{dci}). (c) AC output voltage (V_{ac}).

result, the summit is unavoidable. However, because the peak is only present for a brief time, its impacts are minimal.

The harmonic spectrum of the inverter's output voltage is shown in Figure 9.18. The important harmonics, which are higher-order harmonics, are in the range of 40 to 60. Lower-order harmonics have a magnitude of about 0.05% of the fundamental.

Voltage lift quasi Z-source inverter 239

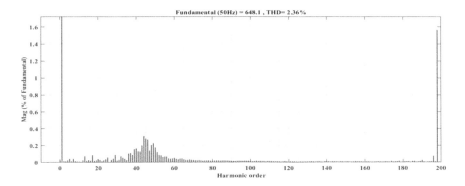

Figure 9.18 THD of output AC voltage.

The input voltage range for the VL-QZSI-enabled application is 150 300 V. The line-to-line rms value of output voltage is 800 V. As a result, peak phase voltage is roughly 653 V. The input voltage (Vdc) of 300 V, the boosted DC voltage (Vdci), the inverter output (Vac) and the output current are shown in Figure 9.19. Peak AC voltage at output end is 642 V. The results for a 150 V input voltage are shown in Figure 9.20. The output voltage peaks at 638.2 V. As a result, even with low input voltages, the inverter generates a good output. As a result, the inverter is considered to have an implied high voltage boost capability.

The inverter's active power output is shown in Figure 9.21. The obtained output power is roughly 72 kW, or 72% of the required power.

Table 9.3 shows the comparison between the proposed topologies.

9.11 Conclusions

In this chapter, an introduction to EVs was provided, and two inverter topologies that are helpful for EVs were discussed. In comparison with the standard ZSI, the topologies provide a good gain. These topologies also have a lower number of switches compared with the other boost-based inverter circuits. They also help in reducing leakage current, as they have dual grounding. Switching schemes corresponding to these topologies were also discussed. The topologies are very suitable for the current trends in the EV domain. Further studies could be carried out on reducing the passive component count used in the circuit, and the present topologies can be made compatible with regenerative braking by making a few modifications in the inverter circuits.

240 *Power electronics for electric vehicles*

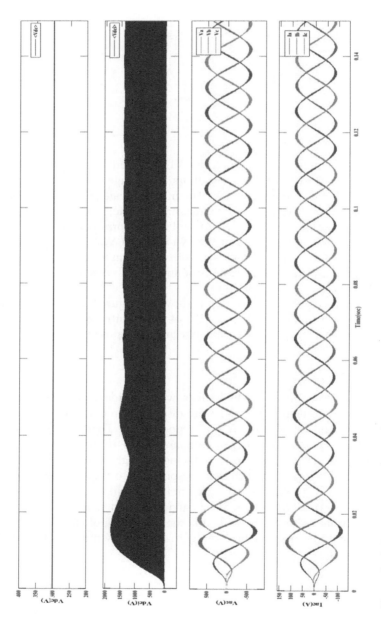

Figure 9.19 For an input voltage of 300 V.

Voltage lift quasi Z-source inverter 241

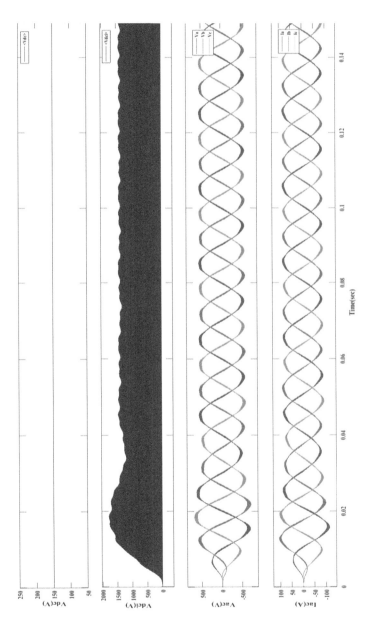

Figure 9.20 For an input voltage of 150 V.

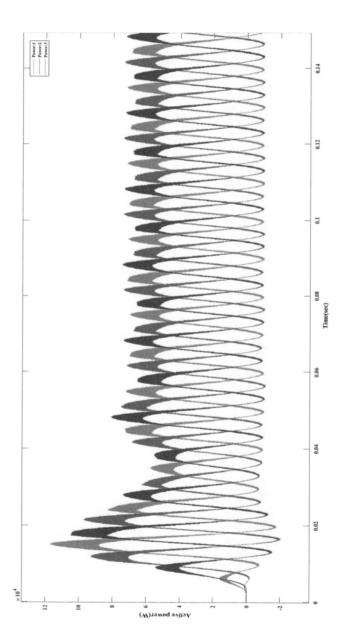

Figure 9.21 Active power output.

Table 9.3 Comparison between the Proposed Topologies

Parameter	Sepic-based inverter	VL-QZSI
Number of phases	1	3
Number of DC links	1	1
Output line voltage	120 V	800 V
Output power	0.5 Hp	100 kW
PWM	Three-state, three-switch-based PWM scheme	Maximum constant boost PWM
Carrier frequency	30kHz	10 kHz
No. of switching devices	3	6
No. of diodes	3	3
No. of passive elements	9	6

References

1. S. Manzetti and F. Mariasiu, "Electric vehicle battery technologies: From present state to future systems," *Renewable and Sustainable Energy Reviews*, vol. 51, pp. 1004–1012, 2015.
2. B. G. Pollet, I. Staffell and J. L. Shang, "Current status of hybrid, battery and fuel cell electric vehicles: From electrochemistry to market prospects," *Electrochimica Acta*, vol. 84, pp. 235–249, 2012.
3. Y. Xue, L. Chang, et al., "Topologies of single-phase inverters for small distributed power generators: An overview," *IEEE Transactions on Power Electronics*, vol. 19, no. 5, pp. 1005–1014, Sept, 2004.
4. Q. Li and P.Wolfs, "A review of the single phase photovoltaic module integrated converter topologies with three different DC link configurations," *IEEE Transactions on Power Electronics*, vol. 23, no. 3, pp. 1020–1033, May 2008.
5. Y. Xue, L. Chang, S. B. Kjær, J. Bordonau and T. Shimizu, "Topologies of single-phase inverters for small distributed power generators: An overview," *IEEE Transactions on Power Electronics*, vol. 19, no. 5, pp. 1005–1014, Sep. 2004.
6. F. Z. Peng, "Z-source inverter," in IEEE Transactions on Industry Applications, vol. 39, no. 2, pp. 504–510, March-April 2003, doi: 10.1109/TIA.2003.808920.
7. P. C. Loh, F. Blaabjerg and C. P.Wong, "Comparative evaluation of pulse width modulation strategies for Z-source neutral-point-clamped inverter," *IEEE Transactions on Power Electronics*, vol. 22, no. 3, pp. 1005–1010, May 2007.
8. V. Vodovozov, N. Lillo and Z. Raud. (2014). Single-phase electric drive for automotive applications. 1313–1318. 10.1109/SPEEDAM.2014.6872106.
9. L. Huang, M. Zhang, L. Hang, W. Yao and Z. Lu, "A family of three-switch three-state single-phase Z -Source Inverters," *IEEE Transactions on Power Electronics*, vol. 28, no. 5, pp. 2317–2329, May 2019.

244 *Power electronics for electric vehicles*

10. J. Anderson and F. Z. Peng, "Four quasi-Z-Source inverters," 2008 IEEE Power Electronics Specialists Conference2008, pp. 2743-2749, J. Anderson and F. Z. Peng, "Four quasi-Z-Source inverters," *2008 IEEE Power Electronics Specialists Conference*, Rhodes, Greece, pp. 2743–2749, 2008.doi: 10.1109/PESC.2008.4592360.

11. Mohamed Ismeil, Kouzou Abdellah, Ralph Kennel, Haitham Abu-Rub and Mohamed Orabi. (2012). "A new switched-inductor quasi-Z-Source inverter topology," 15th International Power Electronics and Motion Control Conference, EPE-PEMC 2012 ECCE Europe Novi Sad, Serbia.

12. T. Takiguchi and H. Koizumi, "Quasi-Z-Source DC-DC converter with voltage-lift technique," *IECON 2010 - 39th Annual Conference of the IEEE Industrial Electronics Society*, Vienna, Austria, pp. 1191–1196, 2013, doi: 10.1109/IECON.2013.6699302.

13. F. Z. Peng, M. Shen and K. Holland, "Application of Z-Source inverter for traction drive of fuel cell—Battery hybrid electric vehicles," *IEEE Transactions on Power Electronics*, vol. 22, no. 3, pp. 1054–1061, May 2007.

14. I.-S. Sorlei, N. Bizon, P. Thounthong, M. Varlam, E. Carcadea, M. Culcer, M. Iliescu and M. Raceanu, "Fuel cell electric vehicles—A brief review of current topologies and energy management strategies," *Energies*, vol. 14, no. 1, p. 252, Jan. 2021.

15. F. Un-Noor, S. Padmanaban, L. Mihet-Popa, M. Mollah and E. Hossain, "A comprehensive study of key electric vehicle (EV) components, technologies, challenges, impacts, and future direction of development," *Energies*, vol. 10, no. 8, p. 1217, Aug. 2017.

16. I. Aghabali, J. Bauman, P. J. Kollmeyer, Y. Wang, B. Bilgin and A. Emadi, "800-V electric vehicle power trains: Review and analysis of benefits, challenges, and future trends," in *IEEE Transactions on Transportation Electrification*, vol. 7, no. 3, pp. 927–948, Sept. 2021.

17. H. N. Tran, T.-T. Le, H. Jeong, S. Kim, H.-P. Kieu and S. Choi, "High power density DC-DC converter for 800V fuel cell electric vehicles," *2021 IEEE 12th Energy Conversion Congress & Exposition - Asia (ECCE-Asia)*, pp. 2224–2228, 2021.

18. Y. Zhang, C. Fu, M. Sumner and P. Wang, "A wide input-voltage range quasi-Z-Source boost DC–DC converter with high-voltage gain for fuel cell vehicles," *IEEE Transactions on Industrial Electronics*, vol. 65, no. 6, pp. 5201–5212, June 2018.

19. Fang Zheng Peng, Miaosen Shen and Zhaoming Qian, "Maximum boost control of the Z-source inverter," *IEEE Transactions on Power Electronics*, vol. 20, no. 4, pp. 833–838, July 2005.

20. M. Shen, Jin Wang, A. Joseph, F. Z. Peng, L. M. Tolbert and D. J. Adams, "Maximum constant boost control of the Z-source inverter," *Conference Record of the 2004 IEEE Industry Applications Conference, 2004. 39th IAS Annual Meeting*, p. 147, 2004.

21. Miaosen Shen, Jin Wang, A. Joseph, Fang Zheng Peng, L. M. Tolbert and D. J. Adams, "Constant boost control of the Z-Source inverter to minimize current ripple and voltage stress," *IEEE Transactions on Industry Applications*, vol. 42, no. 3, pp. 770–778, May–June 2006.

22. Omar Ellabban, Joeri Van Mierlo and Philippe Lataire, "Control of a high-performance Z-Source inverter for fuel cell/supercapacitor hybrid electric vehicles," *World Electric Vehicle Journal*, vol. 4, pp. 444–451, 2011.

10 Sensorless rotor position estimation and regenerative braking capability of a solar-powered electric vehicle driven by PMSM

Vijayapriya R, Arun S L, and Raja Pitchamuthu

10.1 Introduction

The global dependency on energy sources is expanding every day, more specifically in the field of automobiles. The transition is mainly due to rising population, financial changeover, increasing urbanization, the tendency of the public to own their vehicle and upgrading of lifestyle. Eventually, the higher degree of dependency on conventional energy sources such as fossil fuels results in increased environmental pollution by the emission of carbon. It is clear that a significant percentage of carbon dioxide emission is caused by transportation due to the usage of fossil fuels. Most predominantly, road transportation, especially passenger vehicles such as auto-rickshaws, cars and buses, accounts for three-quarters of transport emissions [1]. To ramp up the electrification of transportation units, particularly for two-wheelers, three-wheelers and cars, many start-ups have been initiated by the private and public sectors [2, 3]. This alternative solution is addressed mainly with respect to the range anxiety and high initial cost demanded by internal combustion engine-based vehicles. As battery operated electric vehicles (EVs) are powered from the utility grid, which is the integration of non-renewable energy sources, EVs are not strictly considered as zero-carbon emission vehicles despite the absence of tailpipe emissions. Also, considering the prime concerns of mileage, higher charging time, and the impacts associated with conventional energy sources for the transportation systems, the alternative solution of solar-powered electric vehicles (SPEVs) has been evolved in the market.

Research has been reported in various aspects with respect to solar-powered three-wheeler EVs [4–6]. The usefulness of solar photovoltaic (PV) applications for three-wheeler EVs in the case of the Indian driving system scenario is analysed in Sreejith and Singh [4]. It is proved that a solar PV array provides maximum operating speed at 25 kmh, i.e. 50% of the EV's wheel power. Considering SPEV, either a single- or a double-stage

DOI: 10.1201/9781003248484-10

246 Power electronics for electric vehicles

configuration can be employed to connect the PV array to the powertrain. In this chapter, the double-stage configuration is employed, i.e. a DC–DC boost converter is integrated between the PV array and the DC-link, while the energy storage element battery is directly integrated to the DC-link, avoiding the power conditioning circuit. A boost converter is employed in this work rather than a buck-boost converter due to the high utilization factor, minimum number of switches and passive components, which further result in reduced size and cost [7]. Generally, a permanent magnet brushless DC (PMBLDC) motor is employed as a drive system in battery operated three-wheeler EVs. However, this work employs a permanent magnet synchronous motor (PMSM) due to smooth control, minimal torque ripple, improved efficiency, higher voltage utilization factor and comfortable drive provisioning compared with the PMBLDC motor. Nevertheless, the PMSM drive strategy requires encoders/resolvers for the estimation of rotor position to input the position information to the field-oriented control system. The mechanical sensors involve a complex wiring arrangement; also, they are subjected to temperature variations, which makes the system unreliable. Hence, to enhance the reliability of the EV propulsion system and further, to minimize the cost, research work is focusing more on sensorless estimation of rotor position, avoiding mechanical sensors.

Various sensorless rotor position estimation techniques are addressed in the literature for the PMSM propulsion system [8]. Most prominently, to accurately estimate the high- and low-speed operations, the best ones are reported to be model- and saliency-based approaches, respectively. Accounting for a wider range of operating speed, model-based approaches have been extensively researched for sensorless estimation of rotor position [9–11]. Model reference adaptive control (MRAC) is proven to be simple, robust and efficient; it also provides better performance under variations in the system parameters compared with other computation approaches like sliding mode observer (SMO) and Kalman Filter. Nevertheless, MRAC methods require output filters to suppress the ripples in the computed system parameters such as electromotive force (EMF) and stator flux. The inclusion of filters at the output stage imposes a negative impact on the dynamic response of the models. Also, these conventional methods of rotor information computation result in error amplification due to arctangent operation on estimated EMF and stator flux components. Furthermore, MRAC methods are designed on the basis of numerous assumptions and approximations, and hence, rotor information might not be accurately computed despite lengthy computational algorithms. Inferring that phase locked loop (PLL) is the simplest and most efficient approach to precisely compute the rotor speed and position value, this chapter intends to foremost deploy the basic PLL to compute the rotor information. The objective of employing PLL is to achieve enhanced frequency adaptable capability and quick dynamic response. This is achieved by incorporating pre-stage rotor speed updating and ripple suppression stages. To precisely estimate the rotor information, primarily stator flux needs to be estimated. In this work, the stator flux is directly estimated from the measured stator current and the commanded voltage rather than based on the observer model.

Regenerative braking mechanism is another important aspect to be considered in EVs to achieve increased vehicle range, efficacy and durability [12]. Research works have been addressed to the regenerative braking control strategies for PMSM, such as maximum energy recovery and optimal regulation schemes. An analytical evaluation on the percentage of maximum energy recovered is validated for electric machines through various regenerative braking mechanisms [13–16]. In this chapter, regenerative braking is implemented using the vector control strategies of the machine side converter. Hence, implementing solar-operated EV sensorless estimation of PMSM rotor speed and vector control operation of regenerative braking benefits of the EV are effectively brought out in this work.

The chapter is organized into five sections. The overall modelling of solar PV array, DC–DC converter, voltage source inverter (VSI), PMSM and powertrain stages of the considered system are described in Section 10.2. Section 10.3 presents the proposed control technique for the boost converter and VSI. Also, the sensorless estimation of rotor speed and position is detailed in this section. The analytical design of the overall system performance is discussed in Section 10.4. Section 10.5 concludes the chapter.

10.2 System configuration

The EV system consists of a solar PV, boost converter, battery, DC-link, VSI and three-wheeler EV with the surface-mounted PMSM drive stages as depicted in Figure 10.1. In this work, a solar PV array with an optimum

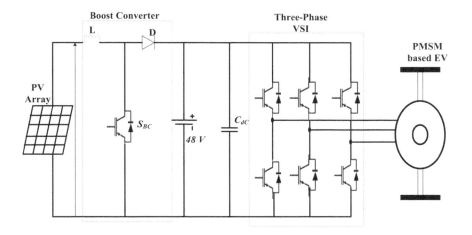

Figure 10.1 Overall system configuration with power conditioning circuits.

248 *Power electronics for electric vehicles*

power rating of 450 W and voltage rating of 48 V is considered. Also, PV voltage regulation using a proportion integral (PI) controller is designed to capture the optimum power, i.e. maximum power point tracking (MPPT) implementation. As illustrated in Figure 10.1, the PV array output terminal is integrated with the EV through the DC–DC converter, DC-link and VSI. The inverter output terminals are connected to PMSM, and the motor drivetrain is coupled with the EV wheels via a gear arrangement to attain the desired speed of the wheel.

10.2.1 *Modelling and selection of solar PV array*

Considering three-wheeler EV size and space, a solar PV array with the optimum power rating of 450 W is used in this work [2]. The analytical modelling of the PV module is tested at the standard conditions with an irradiation and temperature of 1000 W/m^2 and 25°C, respectively. It is also validated that the PV module delivers the rated power of 150 W at the corresponding voltage of 48 V and current of 3.13 A. Since the maximum power rating of 450 W is considered in this work, the PV array needs to be configured with the series and parallel combination of the PV modules. Initially, the voltage corresponding to MPPT is taken to be approximately the same as the DC-link voltage stage, i.e. 48 V [17]. Hence, for the maximum power rating of 450 W, the current corresponding to MPPT is estimated as

$$I_{MPPT} = \frac{P_{PV(MPPT)}}{I_{PV(MPPT)}} \tag{10.1}$$

To match the voltage, current and power ratings, the PV array is configured considering the following combinations:

$$S_{PV(Module)} = \frac{V_{PV(MPPT)}}{V_{(Max)}} \tag{10.2}$$

$$S_{PV(Module)} = \frac{I_{PV(MPPT)}}{I_{(Max)}} \tag{10.3}$$

where $S_{PV(Module)}$ and $P_{PV(Module)}$ represent the number of series and parallel connected modules. $V_{(max)}$ and $I_{(max)}$ represent the maximum voltage and current of a single module. Hence, the PV array configuration considered in this work consists of three parallel connected strings with one PV module in each string. The specifications for the solar PV module and array are given in Table 10.1 in the Appendix. The solar PV array I-V and P-V characteristics for the test condition of 1000 W/m^2 irradiation and 25°C temperature are depicted in Figure 10.2.

Table 10.1 Solar PV Specification

PV Module	
Parameter	Rating
Cells per module	100 V
Open circuit voltage, V_{oc}	62 V
Short circuit current, I_{sc}	3.45 A
Maximum power of the module, PPV(MPPT)	150 W
PV module voltage at MPPT, $V_{PV(MPPT)}$	48 V
PV module current at MPPT, $I_{PV(MPPT)}$	3.13 A
PV Array	
Parameter	Rating
No. of modules connected in series	1
No. of strings connected in parallel	3
Power	450 W

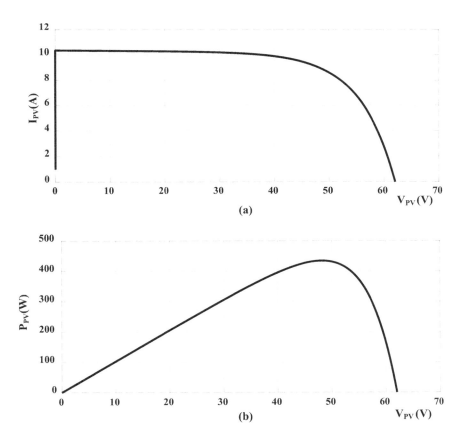

Figure 10.2 Characteristics of solar module: (a) I-V characteristics; (b) P-V characteristics.

250 Power electronics for electric vehicles

10.2.2 Modelling of boost converter

The primary modelling of the DC–DC converter relies on the effective design of the inductor value to ensure continuous current conduction mode, in turn reducing the device stress. Taking into account possible permissible DC-link voltage regulation, the boost converter duty cycle ratio k is calculated as

$$k = \frac{V_{dc} - V_{PV}}{V_{dc}} \tag{10.4}$$

By using the value of k, obtained from Eq. (10.4), switching frequency, f_{sw}, and inductor current ripple, Δ_{iL}, the boost converter inductor value, L, is calculated as given in Eq. (10.5) [18].

$$L = \frac{k V_{PV}}{f_{SW} \Delta_{iL}} \tag{10.5}$$

10.2.3 Modelling of DC-link capacitance

The active power generated by the PV system is supplied to the battery via the boost converter. The battery supplies the DC-link whereby the capacitor C_{dc} is charged from the solar and the electrical machine during normal and regenerative braking conditions, respectively. Assuming the input and output power of the DC-link is equal, i.e. no power loss, the DC-link current can be expressed in terms of converter side i_{dc1} and inverter side i_{dc2} currents as given in Eq. (10.6).

$$C_{dc} \frac{dV_{dc}}{dt} = i_{dc1} - i_{dc2} \tag{10.6}$$

On implementation of sinusoidal pulse width modulation technique on VSI, the relation between the peak value of the fundamental VSI phase voltage and the DC-link voltage can be deduced as given in Eq. (10.7) [19].

$$V = \frac{m V_{dc}}{2\sqrt{2}} \tag{10.7}$$

where m, V and V_{dc} are the modulation index, VSI fundamental phase peak voltage and DC-link voltage, respectively. Generally, DC-link capacitor value is derived on the basis of the admissible percentage of current and voltage ripples in the DC-link. The ripple frequency is subject to variation due to the presence of high-frequency harmonics and electromagnetic interference. The DC-link capacitor value for the allowable deviation in current and voltage can be given as in Eq. (10.8) [18].

$$C_{dc} = \frac{3m\Delta I_{Ripple}}{16\pi f_n \Delta V_{dc}} \tag{10.8}$$

Where f_n, ΔV_{dc} and ΔI_{Ripple} represent the fundamental frequency, voltage ripple and current ripple. It is evident from Eq. (10.8) that, the DC-link voltage ripple can be reduced by increasing the capacitance value. Therefore, the size of the capacitance is decided considering the admissible percentage of voltage ripple. To attain ripple-free sinusoidal output voltage, DC-link voltage ripple is generally considered to be less than or equal to 10% of V_{dc}. Besides, the design of the C_{dc} value is based on the modulation index, which varies according to different operating scenarios. Considering the maximum permissible limit, m is considered to be 1 in this work.

10.2.4 Modelling of three-phase VSI

The three-phase VSI contains a combination of six power switches, i.e. an insulated gate bipolar transistor (IGBT) with an anti-parallel diode. Since the DC-link voltage is maintained constant, i.e. 48 V, the individual IGBT switch voltage rating is calculated as given in Eq. (10.9) with the voltage safety factor, V_{SF}, of 2.

$$V_{IGBT} = V_{SF}V_{dc} \tag{10.9}$$

Also, the individual IGBT switch current rating is obtained considering the safety factor I_{SF} of 1.5 as given in Eq. (10.10).

$$I_{IGBT} = I_{SF}I_{Peak} \tag{10.10}$$

Hence, the power rating of the VSI is obtained by

$$P_{IGBT} = V_{IGBT}I_{IGBT} \tag{10.11}$$

10.2.5 PMSM Modelling

Although a three-phase *abc* sequence-based controller design is feasible, it is far preferable to design the controller with transformation of variables to yield better performance. Transformation denotes change of variables from *abc* to a reference frame (*abc-αβ-dq* frame) that rotates at an arbitrary angular velocity. The requirements of PMSM modelling in a synchronous reference frame (SRF) are described in this section.

(1) Modelling in a-b-c frame

In essence, the circuit model of a PMSM starts with the phase voltage equations in stator coordinates [20]:

$$|U_{abc}| = |i_{abc}||R| + \frac{d|\psi_{abc}|}{dt} \tag{10.12}$$

where the flux linkage ψ_{abc} is given as

$$\psi_{abc} = |L_{abc}(\theta_r)||i_{abc}| + \psi_{fabc}(\theta_r)$$

$$|L_{abc}(\theta_r)| = \begin{vmatrix} L_l + L_0 + L_2\cos 2\theta_r & M_0 + L_2\cos\left(2\theta_r + \frac{2\pi}{3}\right) & M_0 + L_2\cos\left(2\theta_r - \frac{2\pi}{3}\right) \\ M_0 + L_2\cos\left(2\theta_r + \frac{2\pi}{3}\right) & L_l + L_0 + L_2\cos\left(2\theta_r - \frac{2\pi}{3}\right) & M_0 + L_2\cos 2\theta_r \\ M_0 + L_2\cos\left(2\theta_r - \frac{2\pi}{3}\right) & M_0 + L_2\cos 2\theta_r & L_l + L_0 + L_2\cos\left(2\theta_r + \frac{2\pi}{3}\right) \end{vmatrix}$$

where L_l, L_0 and L_2 represent the stator leakage, self and mutual inductance. It is clear from Eq. (10.12) that the phase coordinate inductances are a function of rotor speed, which indicates that the machine inductance values are time-dependent while PMSM is in operation. Hence, the transformation of variables is required to minimize the complexity in the design of the controller. The equivalent circuit of PMSM in stator coordinate ('a-phase') is shown in Figure 10.3. In general, the stator-induced EMF can be expressed as

$$e_j = e_{fj} + e_{ja} + e_{jb} + e_{jc} \tag{10.13}$$

where $j = a$, b and c. e_{fj} represents the permanent magnet (PM)-induced voltage in the corresponding phase, which varies with the rotor position. e_{ja}, e_{jb} and e_{jc} are the self-induced and mutual-motion-induced voltages in the corresponding phase, which are produced by the respective stator currents.

(2) Modelling in α-β frame

The transformation of three-phase AC quantities (*abc*) into the two-phase AC quantities (*α-β*) is generally obtained using Clarke's transformation. The

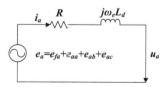

Figure 10.3 Steady-state phase equivalent circuit.

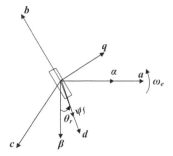

Figure 10.4 PMSM: transformation of variables from *abc* to *dq* frame.

PMSM modelling in the α-β frame, i.e. the stationary reference frame considering counter-clockwise representation of angular speed as depicted in Figure 10.4 [20], can be expressed as

$$u_{s\alpha} = \frac{d\psi_\alpha}{dt} + Ri_\alpha \qquad (10.14)$$

$$u_{s\beta} = \frac{d\psi_\beta}{dt} + Ri_\beta \qquad (10.15)$$

Here, α is assumed to be aligned with 'a-phase', and α-β frame represents the equivalent two-phase alternating current (AC) quantities of *abc* quantities. It is observed that machine inductance in the α-β frame is also time-dependent. Hence, for further simplification, the two-phase AC quantities are transformed into two time-independent quantities, i.e. direct (*d*) and quadrature (*q*), which is defined to be SRF, as depicted in Figure 10.4. The SRF rotates at an arbitrary angular speed ω_e and makes an angle θ_r ($\theta_r = \int \omega_e \, dt$) with the stationary reference frame. Transformation of two-phase AC quantities into two direct current (DC) quantities is obtained using Park's transformation. Complete modelling of PMSM in the *d-q* frame is discussed in the following subsection.

(3) Modelling in d-q frame

According to PM rotor flux alignment as illustrated in Figure 10.4, the machine modelling in the *d-q* frame is given by

$$u_d = -\omega_e \psi_q + Ri_d + \frac{d\psi_d}{dt} \qquad (10.16)$$

$$u_q = \omega_e \psi_d + Ri_q + \frac{d\psi_q}{dt} \qquad (10.17)$$

where ψ_d and ψ_q are the stator flux linkage in the d- and q-axis, which can be defined in terms of stator currents (i_d and i_q), rotor fluxes (ψ_{fd} and ψ_{fq}) and magnetizing inductances (L_d and L_q) as given in Eqs (10.18) and (10.19).

$$\psi_d = L_d i_d + \psi_{fd} \tag{10.18}$$

$$\psi_q = L_q i_q + \psi_{fq} \tag{10.19}$$

Considering the orientation of rotor magnets along the d-axis, the magnetic flux in the q-axis is taken to be zero, and the rotor fluxes in the d-q frame are represented as

$$\psi_{fd} = \psi_{PM} \tag{10.20}$$

$$\psi_{fq} = 0 \tag{10.21}$$

The equivalent circuit of time-independent stator voltages resultant from the transformation of three-phase time-dependent quantities is shown in Figure 10.5. PMSMs are classified as interior- and surface-mounted magnetic machines, and the deduced model is valid for both the machines. However, for interior-mounted magnetic machines, due to saliency in nature, L_d is typically lower than L_q, and the d-q inductances are equal for the surface-mounted magnetic machines, i.e. $L_d = L_q$ [21].

10.2.6 Drivetrain shaft model

The drivetrain shaft of a PMSM-based EV in the one-mass model can be represented as

$$T_e - T_{Lt} = J\frac{d\omega_r}{dt} + B\omega_r \tag{10.22}$$

Where J and B represent the moment of inertia and the frictional coefficient of the mechanical system, respectively. T_m and ω_r are the load torque and the

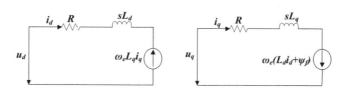

Figure 10.5 PMSM equivalent circuit in SRF.

PMSM rotor mechanical speed, respectively. The machine's electromagnetic torque T_e in terms of pole pair (p_p), rotor flux (ψ_f), stator inductances $(L_d$ and $L_q)$ and currents $(i_d$ and $i_q)$ can be expressed as

$$T_e = 1.5 p_p (\psi_f i_q + (L_d - L_q) i_d i_q) \tag{10.23}$$

In this work, a surface-mounted magnetic machine is employed, which is generally referred to as a non-salient-pole machine. The most important advantages of this machine are simplicity and lower price compared with the interior-mounted, i.e. salient-pole, machine. Therefore, considering non-saliency, Eq. (10.23) can be further deduced, equating $L_d = L_q$. Upon deduction, it can be inferred that T_e depends only on q-axis stator current, and i_q control results in regulation of PMSM active power.

10.3 Control strategies for EV system

The complete control strategy of the EV system considered in this work is depicted in Figure 10.6. Maximum power is extracted from solar, and the boost converter is employed with the corresponding control strategy. Also, one of the primary objectives of VSI control strategies is to achieve maximum torque per ampere, taking into account the driver's demand and the prevailing road profile. To meet this goal, a simple inner current control technique is employed for controlling the switches of the VSI to attain the desired torque and speed target. Additionally, the outer speed control loop

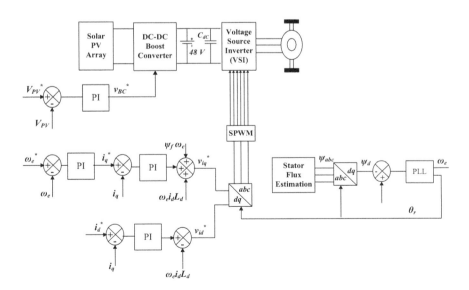

Figure 10.6 Overall controller design for the considered EV system.

employs a sensorless vector control scheme, as depicted in Figure 10.6. The rotor speed is estimated using the PLL technique.

10.3.1 MPPT controller for solar PV array

The energy extracted from solar varies with respect to the prevailing ambient conditions, like sunlight, weather, temperature, irradiation, etc. Hence, to utilize the maximum available solar power under the given scenario and to improve the efficiency and utilization, a simple MPPT control scheme is devised. As the operating point of PV voltage decides the optimum power evacuation from solar, the MPPT controller is framed considering the PV voltage corresponding to maximum power as a reference. The voltage corresponding to the maximum power is computed considering the ambient conditions, i.e. irradiation and temperature. Later, the computed voltage is taken as a set point for the PI regulator, and its output is used to generate the switching pulses for the boost converter, as illustrated in Figure 10.6. The advantages of the proposed technique include simple controller design and avoidance of framing a complex control algorithm.

The boost converter MPPT control technique validation is presented in Section 10.4. The boost converter MPPT control technique updates the duty cycle to continuously provide the switching pulses to the switch, S_{BC}. Power extracted from solar supplies the EV and charges the battery during the motoring and parking mode, respectively. However, in the parking mode, the switching pulses given to the DC–DC converter need to be stopped in case the battery is completely charged, to prevent the battery from overcharging, resulting in temperature increase.

10.3.2 Sensorless speed and position computation using MRAC

As PLL is proven to be an effective method for determining rotor speed and position, in Maiti et al. [22] and Bolognani et al. [23], rotor information is estimated from EMF employing quadrature-PLL, as shown in Figure 10.7.

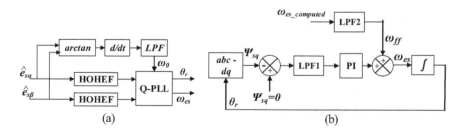

Figure 10.7 Sensorless rotor speed and position computation based on (a) Q-PLL; (b) SRF-PLL – conventional method.

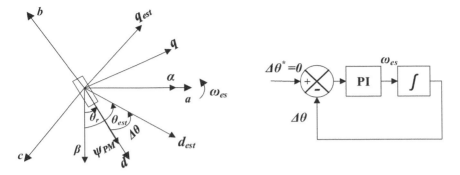

Figure 10.8 Sensorless rotor speed and position computation based on estimated d-q frame – conventional method.

On the contrary, in Park et al. [24] and Hong et al. [25], rotor information is estimated, regulating the rotor position error, $\theta_{r\text{-}Error}$, to be zero via PLL, as illustrated in Figure 10.8. Also, an adaptive PLL technique is discussed in Zdenek and Martin [26] and Ding et al. [27]. In a few of the mentioned methods, $\theta_{r\text{-}Error}$ is derived by subtracting the computed (or observed) value from the measured one. However, in some other methods, $\theta_{r\text{-}Error}$ is computed based on the existence of a coupling component. The existence of a coupling component is identified on one of the estimated axes by imposing a voltage pulse on the other axis. The derived $\theta_{r\text{-}Error}$ is forced to be zero via the PLL controller. Irrespective of the orientation of the estimated d-q-axis and PLL implementation, rotor information estimation is carried out on the basis of model-based techniques. AS MRAC methods are designed on the basis of reduced order calculation with numerous assumptions and approximations, PMSM rotor information may not be accurately computed despite the implementation of a lengthy computational algorithm.

10.3.3 Proposed sensorless speed and position computation

In the proposed method, the rotor angular speed set point is initially determined, and consequently, basic PLL is employed to precisely estimate the rotor information without involving MRAC models. The computational procedure involved in determining the PMSM rotor information is divided into five stages. The first stage is framed to compute the rotor angular frequency set point, while in the second stage, the positive sequence component of flux (d-q) is extracted employing LPF. The third stage integrates the first and second stages by transforming the d-q fluxes to abc using the angle corresponding to the angular speed set point obtained in the first stage. The fourth stage adopts SRF-PLL to compute the stator flux position by

258 *Power electronics for electric vehicles*

orienting the entire stator flux in the assigned dx-axis. Finally, the rotor position information is estimated by finding the association between the rotor and the assigned dx- axis. The detailed procedure of all the computational stages is explained in the following subsections.

(1) PMSM rotor angular speed set point estimation

Stator-induced EMF in the d- and q-axis is given in Eqs (10.24) and (10.25), respectively, as deduced from Eqs (10.16) and (10.17).

$$e_d = \omega_e \psi_q \tag{10. 24}$$

$$e_q = \omega_e \psi_d \tag{10. 25}$$

Rewriting Eqs (2.24) and (2.25), stator fluxes can be expressed with respect to stator current, as given in the following:

$$e_d = \omega_e L_q i_q \tag{10.26}$$

$$e_q = \omega_e L_d i_d + \psi_f \omega_e \tag{10.27}$$

The induced EMF e_q is further deduced as given in Eq. (10.28), since q-axis current is completely confined to electromagnetic torque control, whereas d-axis current is controlled to be zero.

$$e_q = \psi_f \omega_e \tag{10.28}$$

From Eq. (10.28), the rotor angular speed set point ω^*_e can be estimated considering only the positive sequence stator flux and fundamental voltage components.

(2) Rotor information (position and speed) estimation technique

Rotor information (speed and position) is estimated considering additional SRF for the stator flux, i.e. d_x-q_x axes, as depicted in Figure 10.7(a). Stator flux position, δ, with respect to rotor flux can be estimated by orienting the entire stator flux on the d_x-axis such that the q_x-axis stator flux will be forced to be zero. This can be achieved using the SRF-PLL controller, as depicted in Figure 10.7(b). Once the entire stator flux is aligned with the d_x-axis, the rotor position θ_r is determined by deriving the association between the rotor and the defined stator flux d_x-axis.

Dynamic speed tracking and precise position computation are the proven qualities of PLL, provided that the components processed via the loop are free from ripple and the loop is updated with angular frequency. In this context, a feedforward loop with filter [22] and a PLL filter are employed to process the ripple=free components as depicted in Figure 10.7. Generally, filters are employed to filter out the ripple components introduced due to various aspects such as converter non-linearity, harmonics, etc. Moreover, adding filters in the phase detector loop affects the dynamic response in addition to the introduction of complex controller design. To enhance the dynamic behaviour of the PLL, to effectively track the speed and to accurately estimate the position, a pre-stage filter (LPF) and frequency amendment blocks are introduced in this work, as shown in Figure 10.9b. At first, the pre-stage filter is employed to obtain only the positive sequence flux components $\Psi_{dx(PSC)}$, $\Psi_{qx(PSC)}$ from the measured quantities Ψ_{dx}, Ψ_{qx}. Then, the obtained positive DC quantities are converted into abc quantities ($\Psi_{aa(PSC)}$, $\Psi_{ab(PSC)}$ and $\Psi_{ac(PSC)}$) employing the rotor angular frequency set point obtained from the previous step. These abc positive sequence flux components are processed via SRF-PLL to exactly estimate the stator flux angle δ by orienting the entire stator flux Ψ on the dx-axis. Later, rotor position θ_r is determined based on the association between d_f- and d_x-axis as given here:

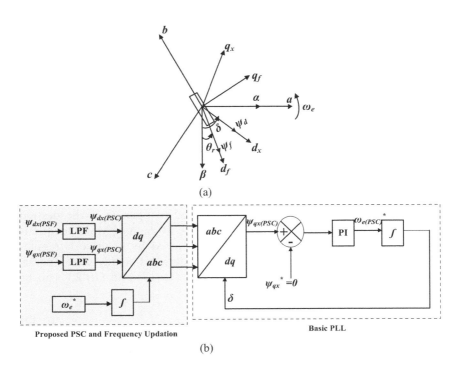

Figure 10.9 PMSM rotor information computation – proposed method. (a) Space vector representation; (b) SRF-PLL.

260 Power electronics for electric vehicles

$$\begin{bmatrix} \Psi_{fd} \\ \Psi_{fq} \end{bmatrix} = \begin{bmatrix} \cos(\delta - \theta_r) & -\sin(\delta - \theta_r) \\ \sin(\delta - \theta_r) & +\cos(\delta - \theta_r) \end{bmatrix} \begin{bmatrix} \Psi_{dx} \\ \Psi_{qx} \end{bmatrix} \tag{10.29}$$

Consistently with rotor flux ($\Psi_{fd} = \Psi_f$, $\Psi_{fq} = 0$) and stator flux ($\Psi_{dx} = \Psi_s$, $\Psi_{qx} = 0$) orientation, Eq. (10.29) can be further deduced, and it can be expressed as

$$\theta_r = \delta - \cos^{-1}\left(\frac{\Psi_f}{\Psi_s}\right) \tag{10.30}$$

$$\theta_r = \delta \tag{10.31}$$

As per the orientation shown in Figure 10.9a, PMSM rotor information is estimated using Eq. (10.30), since θ_r cannot be equal to δ.

Though the proposed method involves PLL and defined d_x-q_x axes to extract rotor information, the method avoids observer models [8, 9, 23] or a reference frame [24–28], as illustrated in Figures 10.7 and 10.8. Furthermore, the technique provides an optimum response with enhanced dynamic behaviour and simple controller design due to the amendment stages of LPF and frequency updating external to the PLL structure. Inconsistently with the traditional approach of rotor information computation by means of PLL, the proposed method is used principally to find the defined stator flux angle, and based on δ, rotor angle is determined. The scheme indicates that complex controller designs such as MRAC methods are not required to estimate the rotor information. This proposed sensorless estimation of rotor information is given as an input to the vector controller of VSI employed in the SPEV system.

10.3.4 Driving and regenerative braking control of PMSM

In normal driving mode, PMSM is controlled to operate at the desired speed as per the controller depicted in Figure 10.6. As PMSM torque is proportional to i_q, the driving torque is achieved by positive i_q regulation of the inner current control loop. Considering a steady-state operating condition and i_d being zero, Eqs (10.16), (10.17) and (10.23) can be rewritten as

$$u_d = -\omega_e L_q i_q \tag{10.32}$$

$$u_q = \omega_e \psi_f + R i_q \tag{10.33}$$

$$T_e = 1.5 p_p (\psi_f i_q) \tag{10.34}$$

Sensorless rotor position estimation 261

Similarly, the regenerative braking mode is achieved by negative regulation of i_q, keeping i_d reference to be zero. The PMSM electromagnetic power, P_e, and stator input power, P_{in}, can be expressed as

$$P_e = \omega_e \psi_f i_{sq} \tag{10.35}$$

$$P_{in} = (\omega_e \psi_f + Ri_{sq})i_{sq} \tag{10.36}$$

During normal driving mode, powers P_e and P_{in} are positive, whereas during regenerative braking mode, both the power values are negative [29].

10.4 Results and discussion

The analytical design of the EV system with the corresponding control scheme is analysed in PSCAD/EMTDC considering 250 μs as a sample time. The specifications of solar PV and PMSM are given in Table 10.1 and Table 10.2, respectively. The validation of the proposed method of sensorless rotor speed and position computation is illustrated in Figure 10.10. Initially, EV is accelerated to 322 rad/sec with a load torque of 100%, i.e. 8 N-m. At 5 s, the vehicle speed is accelerated to 375 rad/sec by keeping the load torque to be 5.8 N-m. It is clear that the proposed system accurately estimates the rotor speed and position, as depicted in Figure 10.10, even during the transient period. Also, it is noted that the error between the estimated value and actual value with respect to speed and position is negligibly small. The speed change is clearly reflected in the position angle, as depicted in Figure 10.10b and c.

The proposed sensorless scheme is also validated by plotting the performance of the VSI's vector control scheme, as depicted in Figure 10.11. The outer speed control loop of the vector control uses the estimated speed to track the reference speed instead of the measured speed. The speed controller generates the corresponding current reference i_{sq}^*, as illustrated in Figure 10.11c. For the corresponding variation in the speed, the controller

Table 10.2 PMSM Specification

Parameter	Rating
Rated Power, P	3000 W
Rated speed, ω_e	375 rad/sec
Rated torque, T	8 N-m
Pole pair, P_p	4
Per phase stator resistance, R_s	0.0085 Ω
Per phase stator inductance, L_s	70 μH
PM flux linkage, ψ_{PM}	0.01641 V-s
Moment of inertia, J	0.012 kg/m^2

262 *Power electronics for electric vehicles*

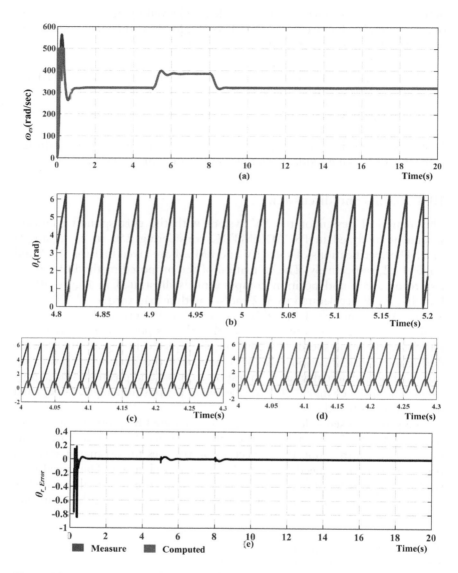

Figure 10.10 PMSM sensorless scheme: (a) computed and measured rotor speeds (rad/s); (b), (c) and (d) computed and measured rotor position (rad) at different speeds; (e) rotor position error.

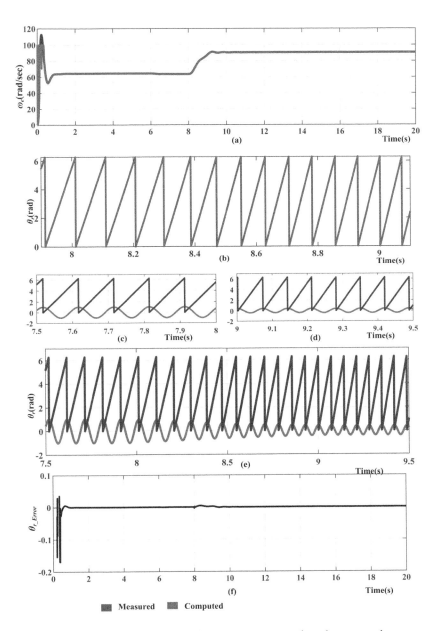

Figure 10.11 PMSM sensorless scheme: (a) computed and measured rotor speed (rad/s); (b), (c), (d) and (e) computed and measured rotor position (rad) at 322 and 375 rad/sec; (f) rotor position error.

264 *Power electronics for electric vehicles*

Table 10.3 Performance Evaluation of Proposed Sensorless Computation

Operating Conditions	Time (s)	ω_{e_Error} (rad/s)	$\theta_{\_Error}$ (rad)
Low Speed	t = 0.01	±5.2	±0.043
	t = 8	±0.083	±0.0006
	t = 9	±0.006	±0.005
	t = 11–18	±0.00009	±0.0003
High Speed	t = 5	±0.078	±0.0013
	t = 6–7	±.0005	±0.0041
	t = 8	±0.07	±0.002

regulates the speed of the PMSM and hence, regulates the stator current. The a-phase stator current is shown in Figure 10.11a for the speed variation of 322 and 375 rad/sec. The reactive power of the PMSM is controlled to be zero, and hence, the current corresponding to its control generates the current $i_d^* = 0$, as depicted in Figure 10.11c. From Figure 10.11e, it is clear that the estimated rotor speed and position accurately match the measured parameters.

The effectiveness of the proposed sensorless scheme is also validated by operating the EV at a lower speed of 90 rad/sec, as depicted in Figure 10.11. Initially, the EV is operated at the speed of 61 rad/sec from time t = 0 s to t = 8 s. From t = 8 s, the speed of the EV is gradually increased, and from t = 9 s, the speed is maintained to be constant at 90 rad/sec. A lower percentage of steady-state and transient error during dynamic operating conditions, as given in Table 10.3, proves the accuracy of the proposed method of PMSM rotor speed and position estimation.

The boost converter is controlled for maximum power extraction under the irradiation of 1000 W/m² from time = 0 to 5 s, and 500 W/m² for the time period from 5 to 10 s. The corresponding PV voltage and current during MPPT are shown in Figure 10.12a and b. The control scheme is also tested under regenerative braking mode. From 0 to 5 s, the battery is charged via the solar and at time = 5 s, the regenerative braking process charges the battery. This is reflected as a battery negative current, as shown in Figure 10.13d. During the entire period, the battery voltage is maintained around 55 V, as depicted in Figure 10.13e.

Data such as solar irradiation and temperature are tracked in the IoT platform ThingSpeak to estimate the maximum voltage, current and power pertaining to the ambient temperature. The data collected from the physical environment, i.e. device layer, is transferred to the cloud, and after predicting the value, it is given in the Simulink environment as an actuation signal.

Sensorless rotor position estimation 265

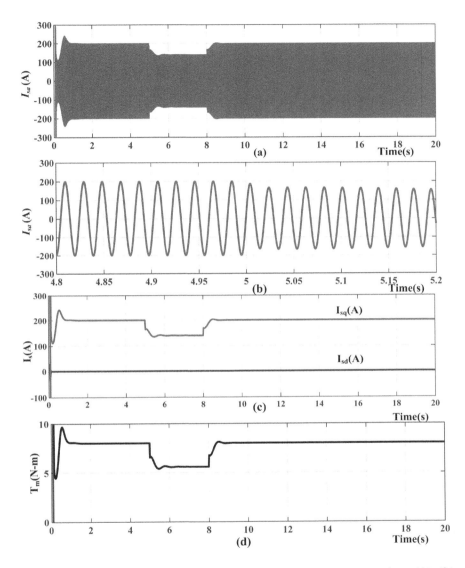

Figure 10.12 PMSM sensorless scheme: (a) PMSM stator current, a-phase (A); (b) zoomed-in view of stator current (a);(c) Stator current in *dq*-axis (A); (d) load torque, N-m.

266 Power electronics for electric vehicles

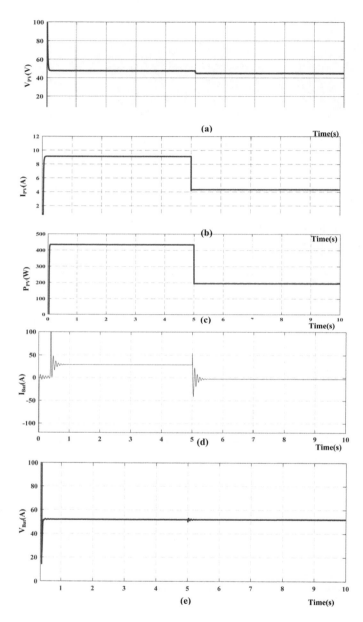

Figure 10.13 EV performance during acceleration and regenerative braking: (a) PV voltage (V); (b) PV current (A); (c) PV power (W); (d) battery current (A); (e) battery voltage during (V).

10.5 Conclusions

A PMSM-driven solar-powered electric vehicle with sensorless rotor position estimation and regenerative braking aspect is analysed in this chapter to analytically investigate the price, range anxiety, efficiency and longevity of three-wheeler EVs. An SPEV driven by PMSM with regenerative braking aspect with respect to the proposed controller is also verified. A simple sensorless rotor position computation model is developed, avoiding the complex observer-based model, and is also proven to be efficient with negligible error. It is evident that the drive reliability can be improved in the absence of dedicated hard-wired speed and position sensors. An optimum torque per ampere regulation of sensorless drive in addition to recovered energy control during regenerative braking is successfully presented using the PSCAD/EMTDC tool. The simplified design of the overall system demonstrates that it is possible to implement the control strategy with the low-cost digital/analog controller. The implemented EV system configuration is a feasible solution for three-wheeler EVs due to various advantages such as enhanced range, low running cost, expanded battery life and less dependence on the utility grid for charging.

References

1. https://worldpopulationreview.com/country-rankings/oil-consumption-by-country.
2. https://ourworldindata.org/co2-emissions-from-transport.
3. https://www.statista.com/statistics/1029864/india-projected-electric-three-wheeler-and-auto-rickshaw-market-size/.
4. R. Sreejith and B. Singh, "Intelligent nonlinear sensorless predictive field oriented control of PMSM drive for three wheeler hybrid solar PV-battery electric vehicle," *2019 IEEE Transportation Electrification Conference & Expo*, Detroit, MI, USA, 2019, pp. 1–6.
5. A. M. Saleque, S. H. Khan, A. M. A. Khan and S. Hoque, "Drivetrain design and feasibility analysis of electric three-wheeler powered by renewable energy sources," *2017 4th International Conference on Advances in Electrical Engineering (ICAEE)*, Dhaka, 2017, pp. 432–438.
6. P. Mulhall, S. M. Lukic, S. G. Wirasingha, Y. Lee and A. Emadi, "Solar-assisted electric auto rickshaw three-wheeler," *IEEE Transactions on Vehicular Technology*, vol. 59, no. 5, pp. 2298–2307, Jun 2010.
7. R. Kumar and B. Singh, "BLDC motor driven water pump fed by solar photovoltaic array using boost converter," *2015 Annual IEEE India Conference (INDICON)*, New Delhi, 2015, pp. 1–6.
8. G. Zhang, G. Wang, D. Xu and N. Zhao, "ADALINE-network-based PLL for position sensorless interior permanent magnet synchronous motor drives," *IEEE Transactions on Power Electronics*, vol. 31(2), 1450–1460, 2016.
9. X. Song, J. Fang, B. Han and S. Zheng, "Adaptive compensation method for high-speed surface PMSM sensorless drives of EMF-based position estimation

268 *Power electronics for electric vehicles*

error," *IEEE Transactions on Power Electronics*, vol. 31(2), 1438–1448, 2016.

10. C. Wu, Y. Zhao and M. Sun, "Enhancing low-speed sensorless control of PMSM using phase voltage measurements and online multiple parameter identification," *IEEE Transactions on Power Electronics*. Early Access. doi: 10.1109/TPEL.2010.2978200

11. Q. An, J. Zhang, Q. An, X. Liu, A. Shamekov and K. Bi, "Frequency adaptive complex-coefficient filter-based enhanced sliding mode observer for sensorless control of permanent magnet synchronous motor drives," *IEEE Transactions on Industry Applications*, vol. 56, no. 1, pp. 335–343, Jan.–Feb. 2010.

12. J. Ko, S. Ko, H. Son, B. Yoo, J. Cheon and H. Kim, "Development of brake system and regenerative braking cooperative control algorithm for automatic-transmission-based hybrid electric vehicles," *IEEE Transactions on Vehicular Technology*, vol. 64, no. 2, pp. 431–440, Feb. 2015.

13. S. Heydari, P. Fajri, M. Rasheduzzaman and R. Sabzehgar "Maximizing regenerative braking energy recovery of electric vehicles through dynamic low-speed cutoff point detection," *IEEE Transactions on Transportation Electrification*, vol. 5, no. 1, pp. 262–270, March 2019.

14. P. Fajri, S. Lee, V. Anand, K. Prabhala and M. Ferdowsi "Modeling and integration of electric vehicle regenerative and friction braking for motor/dynamometer test bench emulation," *IEEE Transactions on Vehicular Technology*, vol. 65, no. 6, pp. 4264–4273, June 2016.

15. S. Murthy, D. P. Magee and D. G. Taylor, "Optimal control of regenerative braking for SPM synchronous machines with current feedback," *2016 IEEE Transportation Electrification Conference and Expo (ITEC)*, Dearborn, MI, 2016, pp. 1–6.

16. K. Choo and C. Won, "Design and analysis of electrical braking torque limit trajectory for regenerative braking in electric vehicles with PMSM drive systems," *IEEE Transactions on Power Electronics,* vol. 35, no. 12, pp. 13308–13321, Dec. 2020 .

17. R. Sreejith and B. Singh, "Position sensorless PMSM drive for solar PVBattery light electric vehicle with regenerative braking capability," *IEEE Energy Conversion Congress and Exposition (ECCE)*, 2010.

18. Xuejun Pei, Wu Zhou and Yong Kang, "Analysis and calculation of DC-link current and voltage ripples for three-phase voltage inverter with unbalanced load," *IEEE Transactions on Power Electronics*, vol. 30, no. 10, pp. 5401–5412, 2015.

19. R. Vijayapriya, P. Raja and M. P. Selvan, "A direct analytical predetermination of PMSG based WPS steady-state values under different operating conditions," *Wind Engineering*, vol.46, no.5, pp. 1570-1589, April 2022.

20. I. Boldea, *The Electric Generators Handbook: Variable Speed Generators*, CRC Press, 2006.

21. Yonggiang Lang Bin Wu, *Navid Zargari and Samir Kouro Power Conversion and Control of Wind Energy System*, IEEE Press, Wiley, 2011.

22. Suman Maiti, Chandan Chakraborty and Sabyasachi Sengupta, "Simulation studies on model reference adaptive controller based speed estimation technique for the vector controlled permanent magnet synchronous motor drive," *Simulation Modelling Practice and Theory*, vol. 17, pp. 585–596, 2009.

23. S. Bolognani, S. Calligaro and R. Petrella, "Design issues and estimation errors analysis of back-EMF-based position and speed observer for SPM synchronous motors," *IEEE Transactions on Power Electronics*, vol. 2(2), 159–170, 2014.

24. Jin-Sik Park, Shin-Myung Jung, Hag-Wone Kim and Myung-Joong Youn, "Design and analysis of position tracking observer based on instantaneous power for sensorless drive of permanent magnet synchronous motor," *IEEE Transactions on Power Electronics*, vol. 27, no. 5, pp. 2585–2594, 2012.

25. Chih-Ming Hong, Chiung Hsing Chen and Chia-Sheng Tu, "Maximum power point tracking-based control algorithm for PMSG wind generation system without mechanical sensors," *Energy Conversion and Management*, vol. 69, pp. 58–67, 2013.

26. N. Zdenek and N. Martin, "Adaptive PLL-based sensorless control for improved dynamics of high-speed PMSM," *IEEE Transactions on Power Electronics* , vol. 37, no. 9, pp. 10154–10165, Sept. 2022.

27. H. C.Ding, X. Zou and J. Li, "Sensorless control strategy of permanent magnet synchronous motor based on fuzzy sliding mode observer," *IEEE Access*, vol. 10, pp. 36743–36752, 2022.

28. L. Tong, X. Zou, S. Feng, Y. Chen, Y. Kang, Q. Huang and Y.Huang, "An SRF-PLL-based sensorless vector control using the predictive deadbeat algorithm for the direct-driven permanent magnet synchronous generator," *IEEE Transactions on Power Electronics*, vol. 29, no. 6, pp. 2837–2849, 2014.

29. X. Zhao, S. Song, Z. Guo and Y. Song, "A regenerative energy management method for permanent magnet synchronous motor control system," *24th International Conference on Electrical Machines and Systems (ICEMS)*, Gyeongju, Korea, Republic of, pp. 2529–2532, 2021.

11 Performance analysis of the integrated dual input converter for EV battery charging application

Kuditi Kamalapathi and
Ponugothu Srinivasa Rao Nayak

11.1 Introduction

The automobile sector is moving towards electrification due to the major impact of transportation through internal combustion engine (ICE) vehicles on global warming. This transition will help to control CO_2 emissions, although it will create transition anxiety for the grid. The rapid transformation of domestic and commercial transport vehicles into a fleet of electric vehicles (EVs) may alter the peak hours of the load profile, which may have major implications for electricity industry operation and planning [1]. This may prove to be a drawback, as the generating stations need to reschedule the units according to changing load profile. Also, charging EVs through electricity from thermal stations will lead to an acceleration of CO_2 emission levels. Thus, for achieving CO_2 emission reduction and reducing dependency on fossil fuels, EVs must be powered through renewable energy stations such as solar and wind [2, 3]. Although such plants need high-accuracy prediction algorithms to schedule generation, their adoption has been rapid, not only in the electrical grid but in various other forms. These include rooftop solar photovoltaic (PV) systems, renewable-powered charging stations and green energy buildings, which have also increased in number due to government policies and customer demand, as they have handsome economic benefits. A charging station powered with energy storage-based solar PV and grid may be an optimal solution [4]. This may reduce the load (EV charger) burden on the grid and meet the EV fleet's charging infrastructure expectations, although it would require a large area of land and additional expensive power electronics to integrate the grid and solar PV power. Also, considering the power demand for EV battery chargers, a solar PV-based charging station alone may not be suitable to support the demand due to intermittency of power. Provision of solar PV power and integration of power from grid and PV can be installed on the EV with the help of dual input DC–DC converters. Although this idea comes with its own difficulties, it eliminates the requirement for PV-based charging stations and off-board power electronics.

DOI: 10.1201/9781003248484-11

272   *Power electronics for electric vehicles*

In on-board solar PV-aided EV charging, the main difficulty is to accommodate the solar panel on the EV's surface [5]. Providing enough surface exposure of solar PV panels to insolation remains the major difficulty. In the literature, solar PV panels in the form of flexible membrane give good efficiency [6]. Also, real implementation and testing of solo solar PV-powered EVs have also been conducted and greatly encouraged by research organizations and institutions for battery charging and motor drives in EVs [7, 8]. One study [9] shows that the driving range of EVs can be extended with on-board solar PV panels. A maximum range extension of 39.9 km was observed for August in Turin under the assumption that the EV runs at 30 km/h. A detailed study of solar-powered EVs is given in Schuss et al. [5] concerning series and parallel connection-type solar PVs, insolation profile for running in urban areas, and variations in surface exposure of the solar PV panel surface due to the shape of the EV's roof. Another major solution provided by solar-powered EVs is the reduction in number of grid-powered EV charging stations, which is analysed in detail in Schuss et al. [5]. It is shown that in countries with yearly low average irradiance as well as high average irradiance, the outcome of investing in solar-powered EVs is positive for the grid, mainly for reducing CO_2 emissions. Thus, with solar power available on the EV's roof, the EV's fleet time at the charging station can be reduced, although this does not eliminate the fact that EVs will stop at a charging station to recharge their battery.

Wireless charging stations are the future of powering EVs, as this makes the charging process fully automated and holds the possibility of charging the vehicles in running condition [10]. A revamp of conventional charging stations to wireless charging stations has received major interest, mainly due to redundancy in station attendees to run the charging stations, automated charging flexibility and the absence of open conductors [11]. Apart from stationary wireless charging, the idea of dynamic charging infrastructure has received major interest, as it extends the range of the EV's journey and provides a solution to range anxiety issues [12, 13]. Another advantage is that with an increase in the rate of charging, the handling of cables becomes difficult, as the size of cables increases, mainly due to heavy current ratings and voltage insulation levels. In wireless charging, the need for handling cables is eliminated irrespective of charger rating; thus, there is no possibility of electrocution from cables. This chapter introduces the integration of solar PV and wireless charging for stationary charging applications. The wireless charger is series-series compensated for 85 kHz standard frequency [14] using spiral circular-shaped coils. To avoid complex and heavy circuitry, series-series compensation is used to achieve resonance condition for wireless power transfer (WPT).

Incorporation of on-board solar PV in a wireless charger-based EV is not seen in the literature and hence, examined in this chapter with the help of DIBBC [15]. Dual-input converters provide solutions for the integration of various sources in EVs, such as supercapacitor and solar PV integration for

Integrated dual input converter 273

battery charging or battery and solar PV integration for powering motor drives, etc [16, 17]. Also, dual-input converters for solar and grid power integration are seen in the literature [18]. The design and selection of suitable dual-input converter topologies depends on the type, the ratings of the sources being integrated and the load. Further, designing a controller for such converter topologies is another difficult task [19, 20]. The integration of solar PV and grid power will require the solar PV power to be extracted under maximum power point tracking (MPPT) conditions, and thus, the control should be based on an integrated MPPT scheme [21]. It can prove to be an complex and exhausting task to design and optimize such a control scheme. Thus, in this chapter, the two sources, wireless charger and solar PV are controlled using a conventional PI controller. The intermittencies in output power of solar PV are assumed to be zero in this chapter to test the converter with a conventional PI controller. The controller is designed from the transfer function of the DIBBC derived from its small-signal model [22, 23]. Simulation and hardware results are both given for a clear explanation of the integrated charging system analysis.

An on-board solar PV and WPT integrated EV charging system has been simulated, and a hardware prototype for the same is developed and tested. Section 11.2 focuses on the architecture of the designed charging system and design equations of the two power sources. In Sections 11.3 and 11.4, modes of operation and small-signal model of the non-ideal DIBBC are derived and used to derive the transfer function for controlling the switches in different modes of operation. Section 11.5 discusses the control scheme and controller design for the solar PV and wireless power integrated DIBBC. In Section 11.6, simulation results, details of hardware prototype and hardware results are discussed. Also, the merits and demerits of the proposed EV charging system are addressed in this section. The conclusion is given in Section 11.7.

11.2 System description

The proposed charging system architecture is shown in Figure 11.1. The main parts of the dual-input charging system are wireless charger, solar PV array, controller, battery and DIBBC. The diagram includes a solar PV array and a detailed wireless charger circuit. Three switches, S_1, S_2 and S_3, are connected between the source section and converter output through single inductor L and diode D. The solar PV source consists of the solar panel and reverse protection diode D_{PV}. The wireless charger system is powered by a direct current (DC) source, as in the proposed architecture, or it can be powered from the grid through a rectifier and filter, such as in a battery charging station. The power from the DC source is converted to high-frequency alternating current (AC), i.e. 85 kHz, through a high-frequency inverter. This high-frequency AC powers a primary or transmitter coil network consisting of an inductor L_1 and a compensation capacitor C_1, designed to achieve series resonance condition at 85 kHz frequency. Similarly, the same

274 Power electronics for electric vehicles

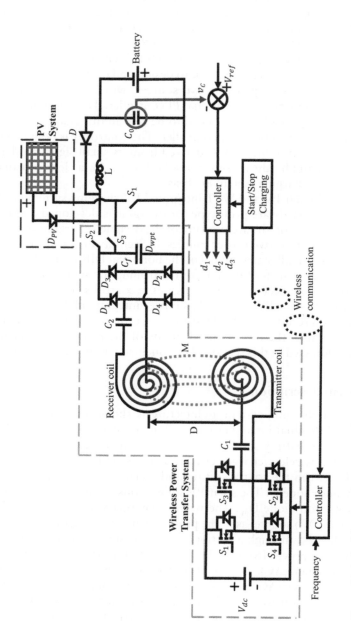

Figure 11.1 Architecture for Integrated WPT plus solar PV based EV charging.

values of inductor and capacitor are used to design a secondary or receiver side series resonant network for 85 kHz frequency. Across the receiver side output terminals, a rectifier is connected, which converts the received AC power to DC. A constant output voltage is maintained at one input of the dual-input DC–DC converter under no-load conditions. Solar PV acts as the second input to the dual-input converter. The working of the DIBBC is discussed in detail in Section 11.3.

11.2.1 On-board PV system

An accurate equivalent circuit of the solar PV source is shown in Figure 11.2. In the proposed architecture, the solar panel needs to be placed on the roof top of the EV. This will allow regular cooling of the panel while driving and charging the battery. Under favourable weather conditions, the photo current generated by the panel is I_{ph} and is computed using *Eq. (11.1)*. Current supplied by the PV panel to the dual input converter is given by *Eq. (11.2)*. Solar PV current and voltage are denoted by I_{PV} and V_{PV}. Energy from solar irradiance (I_r) is converted to photo current under temperature (T). The short circuit current (I_{sc}) under 1000 W/m² is given by K_i, and R_{sh}, R_s are the internal shunt and series resistance of the PV panel. Thermal voltage (V_t) across the shunt diode(D) varies with temperature. Based on load requirement, the number of series (N_s) and parallel (N_p) connected solar PV modules can be selected to get the desired output voltage and current. The saturation current (I_0) of the solar PV panel varies with the number of photons and frequency of irradiance shown on the PV panel.

$$I_{ph} = \left[I_{sc} + K_i (T - 298) \right] \times \frac{I_r}{1000 I_r} \tag{11.1}$$

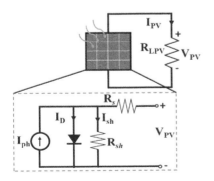

Figure 11.2 Solar PV equivalent circuit.

$$I_{PV} = N_P I_{ph} - N_P I_0 \left[\exp^{\frac{\frac{V_{PV}}{N_S} + \frac{I_{PV} R_s}{N_P}}{nV_t}} - 1 \right] - I_{sh} \quad (11.2)$$

The open-circuit voltage (V_{oc}) can be derived from *Eq. (11.2)* by substituting I = 0 and short circuit current (I_{sc}) by substituting V = 0. The voltage and current supplied by the solar PV module to load are V_{PV} and I_{PV}. Discussion of the solar PV system is restricted to basic working.

11.2.2 Wireless power transfer system

The equivalent circuit of the two-coil wireless charger portion of the proposed system is shown in Figure 11.3.

In Figure 11.3, Vs and I_1 are the primary source voltage and current. V_{WPT} is the receiver side voltage across the load R_{LW}, and current across the load on the receiver side is I_2. R_g is the source resistance on the primary side. R_T and R_R are the transmitter coil and receiver coil resistance. Compensating capacitors C_T and C_R are required to compensate the coil inductance L_T and L_R to obtain the resonance condition in a WPT system. The mutual inductance of the two-coil system between the transmitter coil (TxC) and receiver coil (RxC), (M_{TR} or M_{RT}) is denoted by M.

R_{LW}, the load resistance referred to the secondary side from the rectifier output, is given by

$$R_{LW} = \left(\frac{8}{\pi^2}\right)(R_L) \quad (11.3)$$

The reflected resistance (R_{re}) from the receiver side to the transmitter side is given by

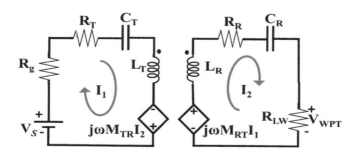

Figure 11.3 WPT equivalent circuit.

$$R_{re} = \frac{(\omega M)^2}{R_R + R_{LW}} \tag{11.4}$$

The total resistance (R_{Tot}) of the two-coil circuit referred to the primary is given by

$$R_{Tot} = R_T + R_g + R_{re} \tag{11.5}$$

The relation is derived by taking consideration of the resonance condition, i.e., inductive reactance $(X_L = X_C)$.

$$\text{Resonance frequency}\,(\omega) = \frac{1}{\sqrt{L_T C_T}}\ \text{rads/sec} \tag{11.6}$$

By applying Kirchhoff's voltage law (KVL) to the circuit,

$$V_S = I_1(R_g + R_T) - j\omega M I_2 \tag{11.7}$$

$$j\omega M I_1 = I_2(R_R + R_{LW}) \tag{11.8}$$

The primary coil and secondary coil currents are as given by

$$\begin{bmatrix} I_1 \\ I_2 \end{bmatrix} = \frac{1}{\left\{ j\omega M^2 - \left(R_g + R\right)\left(R_R + R_{LW}\right)\right\}}$$

$$\begin{bmatrix} -\left(R_R + R_{LW}\right) & -j\omega M \\ -j\omega M & R_g + R_T \end{bmatrix} \begin{bmatrix} V_S \\ 0 \end{bmatrix} \tag{11.9}$$

The input and output power of the two-coil system is given by

$$P_{in} = \left|I_1\right|^2 \left(R_g + R_T + R_{re}\right) \text{and} P_{out} = \left|I_2\right|^2 \left(R_{LW} R_{re}\right) \tag{11.10}$$

The simplified expression for the efficiency of the two-coil system is given by

$$\eta = \frac{1}{a R_{LW} + \left(\dfrac{\beta}{R_L}\right) + \gamma} \tag{11.11}$$

In Eq. (11.11), , and are given by

$$a = \frac{R_g + R_R}{(\omega M)^2}; \beta = \left(\left(\left(R_g + R_T\right)\frac{R_r}{(\omega M)^2}\right) + 1\right) R_T; \gamma = \left(2\left(R_g + R_T\right)\frac{R_R}{(\omega M)^2}\right) + 1$$

11.2.3 Design equations of inductance, capacitance of coils

In this chapter, an amended form of the original Wheeler formula for a spiral circular coil (shown in Figure 11.4) has been used to estimate the value of self-inductance.

$$L(\mu H) = \frac{N^2 R^2}{(8R + 11W)} \tag{11.12}$$

where
  N = number of turns
  $R = 0.5(r_{out} + r_{in})$ = mean radius of the spiral coil
  $W = (r_{out} - r_{in})$ = depth of the spiral coil

The expression for the total length of the spiral circular coil is $l = \pi N(r_{out} + r_{in})$, where r_o and r_{in} are the outer and inner radius of the spiral circular coil. The minimum and maximum values of r_o can be calculated as

$$r_{out}(\max) = r_{in} + N*(D+S) \tag{11.13}$$

$$r_{out}(\min) = r_{in} + N*D \tag{11.14}$$

The DC and AC resistance of the spiral circular multi-strand Litz wire is respectively given as follows:

$$R_{dc} = \frac{4\rho_c l_{tot}}{\pi n d_s^2} \tag{11.15}$$

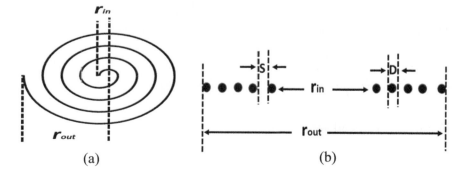

Figure 11.4 Representation of a spiral circular coil. (a) Three-dimensional view of the spiral circular coil; (b) cross-sectional view of the spiral circular coil.

$$R_{ac} = \frac{384(8\pi f 10^{-7})R_{dc}^3}{\left((8\pi f 10^{-7})^2 + R_{dc}^2\right)^2} \tag{11.16}$$

Where n is number of strands, ρ_c is the resistivity of the conducting material, d_s is the diameter of the strand, l_{tot} is the total length of the spiral coil and f is the operating frequency of the coil.

11.3 Modes of operation

The circuit diagram of the DIBBC is shown in Figure 11.5. V_{PV} and V_{WPT} represent the voltage values of the two sources, and S_1, S_2 and S_3 are the three switches of the DIBBC. The remaining part of the DIBBC is similar in design to a conventional buck-boost converter with inductor (L), capacitor (C) and load resistance R_L. The working of the dual input converter is shown in Figure 11.6a–d. Table 11.1 provides the details for state of switches, diodes and other elements and load in the circuit. The timing diagram in Figure 11.7 shows states of the gate pulses given to the switches along with voltage and current across the inductor.

Mode 1: In this mode, switch S_1 is ON. The energy is delivered to the inductor from the source V_{PV}, as shown in Figure 11.6a. Here, the diode D and switches S_2 and S_3 are in a non-conduction state.

Mode 2: In this mode of operation, switch S_3 is turned ON, as shown in Figure 11.6b. From the figure, it is clear that the conduction of switch S_3 helps to make a series combination of both input sources together. So, in this particular mode of operation, energy is delivered to the inductor simultaneously from both the input sources. This is one of the potential merits of the proposed converter, in which the individual and simultaneous utilization of both input sources is possible. This unique operation of the proposed converter results in potential applications like a hybrid EV, renewable energy integration, etc. In this mode, switches S_1, S_2 and diode D are in the non-conduction state.

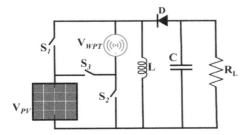

Figure 11.5 Dual-input buck-boost DC–DC converter.

280 Power electronics for electric vehicles

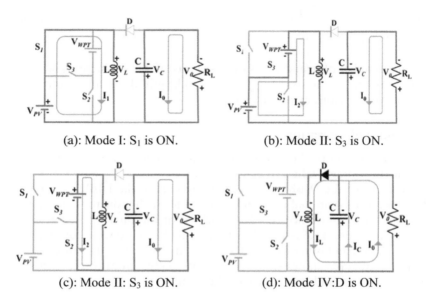

Figure 11.6 (a) Mode I: S_1 is ON. (b) Mode II: S_3 is ON. (c) Mode II: S_3 is ON. (d) Mode IV:D is ON.

Table 11.1 Modes of Operation

Mode	S_1	S_2	S_3	D	V_L	I_L	V_0	I_0
I (t = 0 to t = d_1T_s)	ON	OFF	OFF	OFF	V_{PV}	I_1	$-V_C$	I_C
II (t = d_1T_s to t = d_2T_s)	OFF	OFF	ON	OFF	$V_{PV}+V_{WPT}$	I_3	$-V_C$	I_C
III (t = d_2T_s to t = d_3T_s)	OFF	ON	OFF	OFF	V_{WPT}	I_2	$-V_C$	I_C
IV (t = d_3T_s to t = T_s)	OFF	OFF	OFF	ON	$-V_0$	I_L	V_C	I_0

Mode 3: Here, the switch S_2 alone is ON, while the remaining switches are in OFF state. In this mode, energy is delivered to the inductor from the second input source V_{WPT}, as given in Figure 11.6c.

Mode 4: In this mode, all the switches S_1, S_2 and S_3 are in OFF state, and the diode D becomes forward biased. Hence, the energy stored in the inductor L is delivered to the load and also used to charge the capacitor C at the output side, as given in Figure 11.6.

When the suggested converter is employed in an EV application, it is possible for the drive motor to generate power while braking. The proposed

Integrated dual input converter 281

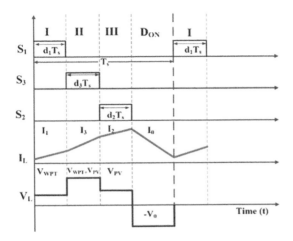

Figure 11.7 Timing diagram for continuous-conduction mode (CCM).

converter has the capability of recovering energy from the load and charging the input sources (Figure 11.7).

11.4 Modelling of DIBBC

The state-space model of DIBBC can be expressed as given in Eqs (11.17) and (11.18).

$$[M]\dot{X} = AX + BU \qquad (11.17)$$

$$Y = CX + DU \qquad (11.18)$$

Here, A is the state matrix, B is the input matrix, C is the output matrix and D is the feed-forward matrix. Matrix $\dot{X}$, U, Y and [M] are as given here:

$$\dot{X} = \begin{bmatrix} \dot{i}_L \\ \dot{V}_C \end{bmatrix}; M = \begin{bmatrix} L & 0 \\ O & C \end{bmatrix}; Y = \begin{bmatrix} V_O \\ I_1 \\ I_2 \\ I_3 \end{bmatrix}; U = \begin{bmatrix} V_{PV} \\ V_{WPT} \end{bmatrix}$$

The small-signal model of the DIBBC is derived and given by Eqs (11.19) and (11.20).

282 *Power electronics for electric vehicles*

$$\begin{bmatrix} L & 0 \\ 0 & C \end{bmatrix}\begin{bmatrix} \tilde{i}_L \\ \tilde{v}_c \end{bmatrix} = \begin{bmatrix} 0 & -1+\left(d_1+d_2+d_3\right) \\ 1-\left(d_1+d_2+d_3\right) & \dfrac{-1}{R} \end{bmatrix}\begin{bmatrix} \tilde{i}_L \\ \tilde{v}_c \end{bmatrix}$$

$$+\begin{bmatrix} d_1+d_3 & d_2+d_3 \\ O & 0 \end{bmatrix}\begin{bmatrix} \tilde{v}_{PV} \\ \tilde{v}_{WPT} \end{bmatrix}+\begin{bmatrix} V_{PV}+V_C \\ -I_L \end{bmatrix}\tilde{d}_1 \qquad (11.19)$$

$$+\begin{bmatrix} V_{WPT}+V_C \\ -I_L \end{bmatrix}\tilde{d}_2+\begin{bmatrix} V_{PV}+V_{WPT}+V_C \\ -I_L \end{bmatrix}\tilde{d}_3$$

$$\begin{bmatrix} v_O \\ i_1 \\ i_2 \\ i_3 \end{bmatrix} = \begin{bmatrix} 0 & -1 \\ d_1 & 0 \\ d_2 & 0 \\ d_3 & 0 \end{bmatrix}\begin{bmatrix} \tilde{i}_L \\ \tilde{v}_c \end{bmatrix}+\begin{bmatrix} 0 & 0 \\ 0 & 0 \\ 0 & 0 \\ 0 & 0 \end{bmatrix}\begin{bmatrix} \tilde{v}_{PV} \\ \tilde{v}_{WPT} \end{bmatrix}+\begin{bmatrix} 0 \\ I_L \\ 0 \\ 0 \end{bmatrix}\tilde{d}_1$$

$$+\begin{bmatrix} 0 \\ 0 \\ I_L \\ 0 \end{bmatrix}\tilde{d}_2+\begin{bmatrix} 0 \\ 0 \\ 0 \\ I_L \end{bmatrix}\tilde{d}_3 \qquad (11.20)$$

To obtain the small-signal model circuit, expand the state equation (11.19):

$$L\frac{dI_L}{dt} = -\left(1-d_1-d_2-d_3\right)\tilde{v}_c+(d_1+d_3)V_{PV}+(d_2+d_3)V_{WPT}$$

$$+\left(V_{PV}+V_C\right)\tilde{d}_1+\left(V_{WPT}+V_C\right)\tilde{d}_2+\left(V_{PV}+V_{WPT}+V_C\right)\tilde{d}_3 \qquad (11.21)$$

$$C\frac{dV_C}{dt} = \left(1-d_1-d_2-d_3\right)\tilde{i}_L-\frac{1}{R}\tilde{v}_c-I_L\tilde{d}_1-I_L\tilde{d}_2-I_L\tilde{d}_3 \qquad (11.22)$$

Also, from output equation (11.20):

$$v_O = -v_c, i_1 = d_1\tilde{i}_L+I_L\tilde{d}_1, i_2 = d_2\tilde{i}_L+I_L\tilde{d}_2, i_3 = d_2\tilde{i}_L+I_L\tilde{d}_3 \qquad (11.23)$$

By using the equations derived from Eqs (11.21) and (11.22), the equivalent small-signal modal circuit is obtained for the three different duty ratios shown in Figure 11.8.

Integrated dual input converter 283

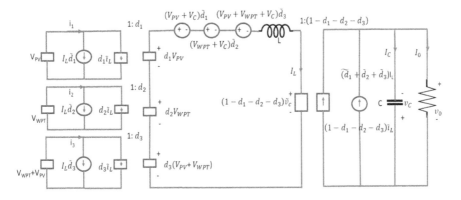

Figure 11.8 Small-signal model circuit of DIBBC.

The effect of small-signal test disturbance can be observed through analysing the transfer functions derived from the small-signal model. The response of $\tilde{v}_c$ due to small-signal test disturbance injected in the duty ratio $\tilde{d}_1$ can be measured based on the small-signal model given in Eqs (11.14) and (11.15) by fixing the remaining input small-signal test disturbance values $\tilde{v}_{s1}, \tilde{v}_{s2}, \tilde{d}_2$ and $\tilde{d}_3$ to zero. Similarly, the response of $\tilde{v}_c$ due to small-signal test disturbances in $\tilde{d}_2$ and $\tilde{d}_3$ can be measured with the transfer functions given in Eqs (11.24), (11.25) and (11.26).

$$\frac{\tilde{v}_c(s)}{\tilde{d}_1(s)} = \frac{[1-(d_1+d_2+d_3)](V_{PV}+V_C)-SLI_L}{S^2LC+S\frac{L}{R}+[1-(d_1+d_2+d_3)]^2} \tag{11.24}$$

$$\frac{\tilde{v}_c(s)}{\tilde{d}_2(s)} = \frac{[1-(d_1+d_2+d_3)](V_{WPT}+V_C)-SLI_L}{S^2LC+S\frac{L}{R}+[1-(d_1+d_2+d_3)]^2} \tag{11.25}$$

$$\frac{\tilde{v}_c(s)}{\tilde{d}_3(s)} = \frac{[1-(d_1+d_2+d_3)](V_{PV}+V_{WPT}+V_C)-SLI_L}{S^2LC+S\frac{L}{R}+[1-(d_1+d_2+d_3)]^2} \tag{11.26}$$

The three transfer functions (11.24), (11.25) and (11.26) are evaluated with the respective step response shown in Figure 11.9. The step response settles at their respective DC gain values of 507, 580.76 and 691.1. Peak overshoot of the three transfer functions varies, as the input voltage for each transfer function is different. Solar PV and WPT source voltages are 60 V and 36 V. It can be seen that the peak overshoot response and steady-state value for the

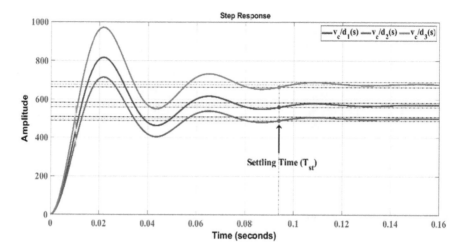

Figure 11.9 Step response of transfer functions.

transfer function (11.21) are higher, as the input voltage and DC gain are higher compared with the transfer function (11.19) and (11.20). Thus, peak overshoot magnitude and oscillations depend on the DC gain and input voltage. Settling time (T_{st}) for the three transfer functions is equal to 0.095 seconds as obtained from the step response. The pole-zero map and Bode plot shown in Figures 11.10 and 11.11 indicate that the system is stable, as there

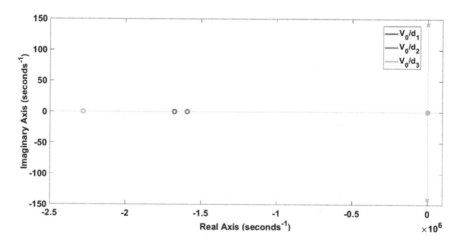

Figure 11.10 Pole-zero location of transfer function.

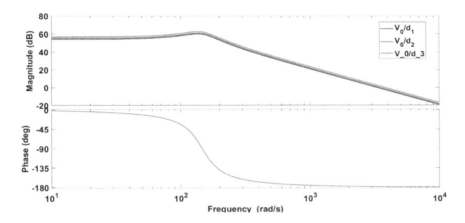

Figure 11.11 Bode plots of open-loop transfer function.

are no open loop poles on the RHS of the S-plane, and the phase margin of the three transfer functions is positive. The eigen values of the characteristic equations are $-0.3846 + 1.4109i$ and $-0.3846 - 1.4109i$. Also, it can be seen that the peak overshoot and DC gain of (11.21) are higher, so the stability limit, i.e., the phase margin, is lowest for $\dfrac{\tilde{v}_c(s)}{\tilde{d}_3(s)}$. A suitable controller is thus required to shift the phase margin of the transfer functions to greater than at least 40 dB.

11.5 Control scheme for the DIBBC

The controller for the DIBBC is modelled as shown in Figure 11.12. The error signal E is fed to three PI controllers. The value of K_p and K_i for each controller is modelled using the transfer functions (11.19), (11.20) and (11.21). Switching time d_1T_s, d_2T_s and d_3T_s for the switches S_1, S_2 and S_3 is controlled by converting the output of PI controllers to time delay ($T_{\delta 1}$, $T_{\delta 2}$ and $T_{\delta 3}$) values for the respective switching time period. This time delay value is used to delay the reference sawtooth carrier of the next switch being turned ON. Switch S_1 is set as the reference, i.e., the sawtooth carrier for S_1 is not delayed, and P_1I_1 output is used to delay the S_3 switch carrier. The addition of P_1I_1 and P_2I_2 is used to delay the carrier of S_2. The output of P_3I_3 is not used to delay, as there is no next switch for controlling. This scheme can be adopted for N input DC–DC converter as well.

Bode plots for compensated and uncompensated systems are shown in Figure 11.13. The details of gain margin (GM) and phase margin (PM) of

286 Power electronics for electric vehicles

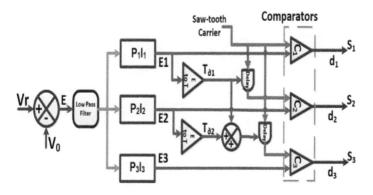

Figure 11.12 Control scheme for DIBBC.

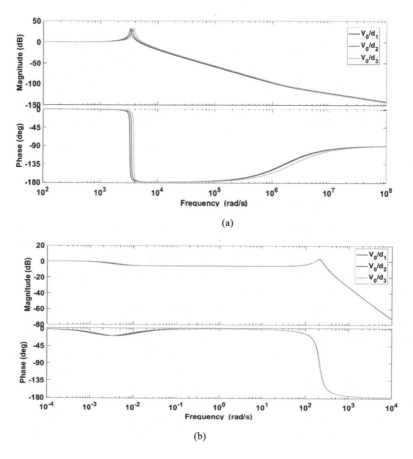

Figure 11.13 Closed-loop Bode plots: (a) without controller, (b) with controller.

Integrated dual input converter 287

Table 11.2 Relative Stability Parameters

Transfer function	Gain margin (GM)	Phase margin (PM)	Stability
$\dfrac{V_0}{d_1} = \dfrac{6.4809(s+1.675e^{06})}{s^2+76.92s+2.139e^{04}}$	infinity	1.47	Stable
$\dfrac{V_0}{d_2} = \dfrac{7.808(s+1.591e^{06})}{s^2+76.92s+2.139e^{04}}$	infinity	1.37	Stable
$\dfrac{V_0}{d_3} = \dfrac{6.4809(s+2.281e^{06})}{s^2+76.92s+2.139e^{04}}$	infinity	1.25	Stable

Table 11.3 Closed-Loop Stability Analysis of DIBB Converter

Transfer function	K_p	K_i	Controller transfer function	GM and PM with controller	W_{gc} and W_{pc}
$\dfrac{V_0}{d_1}$	0.0029	0.0001	$\dfrac{0.0029s+0.0001}{s}$	42, Infinity	Infinity, 276.04
$\dfrac{V_0}{d_2}$	0.0027	0.0001	$\dfrac{0.0027s+0.0001}{s}$	40.1, Infinity	Infinity, 283.7
$\dfrac{V_0}{d_3}$	0.0022	0.0001	$\dfrac{0.0022s+0.0001}{s}$	41.4, Infinity	Infinity, 279.7

the controller transfer function with gain cross over frequency (W_{gc}) and phase cross over frequency(W_{pc}) are given in Table 11.2.

11.6 Results and discussion

11.6.1 *Simulation results*

The DIBBC design is simulated in MATLAB Simulink. PI controller values designed with the help of transfer functions

$$\frac{\tilde{v}_c(s)}{\tilde{d}_1(s)}, \frac{\tilde{v}_c(s)}{\tilde{d}_2(s)} \text{ and } \frac{\tilde{v}_c(s)}{\tilde{d}_3(s)}$$

are used in the designed control scheme explained in the previous section. The steady-state switching pulses and inductor voltage for the four modes of operation observed in the simulation are shown in Figure 11.14a. The switching pattern of the switches matches well with the theoretical timing diagram. Inductor current and capacitor current ripples under steady state are shown in Figure 11.14b and c. In the rising portion of the inductor current (I_L), three slopes are observed, which indicate the initial three modes

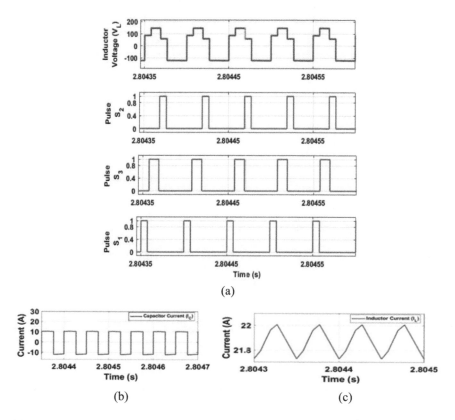

Figure 11.14 DIBBC behaviour: (a) inductor voltage and switching pulses; (b) inductor current; (c) capacitor current (I_C).

of operation during which the inductor is charged from the sources. The capacitor current (I_C) profile consists of charging and discharging current. The negative current is the discharging portion, i.e., the first three modes of operation of the dual-input converter. The positive portion is the charging operation during the fourth mode of operation of the converter. During the fourth mode, the inductor discharges across the capacitor along with the load to power up the capacitor for the next switching cycle. Output voltage ripple of 1 V is observed in the voltage profile during the steady state Figure 11.13. This is less than 1% of the output voltage. Thus, the inductor and capacitor values are well suitable for the simulated source voltages and desired output voltage of 130 V and 10 A load current.

Controller performance is evaluated by injecting a disturbance into the two DC sources. The output voltage and output current during the disturbance injection into the sources are shown in Figure 11.15a and b,

Integrated dual input converter 289

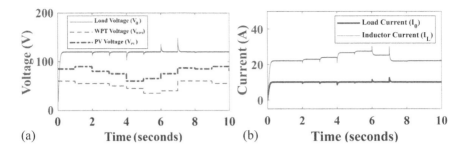

Figure 11.15 Effect of input source variations: (a) system voltages, (b) load current.

respectively. In the initial state of the output voltage, the output current is settled at 130 V/10 A. During the disturbance injection into both sources, the output voltage and current deviate for a short time from the desired values. When both the sources have negative disturbance, the output voltage falls, and vice versa. The effect on output voltage is minor if one input source suffers positive disturbance while the other suffers negative disturbance. In all the disturbances shown in Figure 11.14a, the controller is robust enough to maintain the output voltage equal to the desired value within a short period after the disturbance injection. During the disturbance, the inductor current reduces, increases when the sum of the voltage sources deviates in positive, negative from the normal condition (60 V + 36 V).

11.6.2 Hardware results

The hardware setup shown in Figure 11.16a is tested for the desired output voltage of 130 V with resistive load and a battery stack, shown in

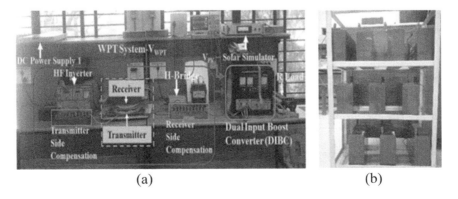

Figure 11.16 Hardware setup: (a) integrated wireless plus solar PV based EV charging system test bench, (b) battery stack.

290 *Power electronics for electric vehicles*

Figure 11.16b. Details of different specifications of the hardware setup are given in Tables 11.4, 11.5 and 11.6. Solar PV is substituted by a solar simulator for hardware analysis and evaluation of the working of the proposed system. The high-frequency inverter and DIBBC are independent controlled switching pulses by a SPARTAN-6 FPGA board, shown in Figure 11.17.

The coil arrangement for the WPT system is shown in Figure 11.18a. Before being connected to the DIBB converter, the WPT system is tested for the different distances to obtain the optimal distance; a distance of 10 cm distance is chosen because of its average good efficiency and also better output power. The efficiency and power plots for various distances between the transmitter and the receiver are shown in Figure 11.18b and c. The transmitter coil and

Table 11.4 Wireless Power Transfer (WPT) System Specifications

Parameter	Transmitter coil (TxC)	Receiver coil (RxC)
Turns	27	27
Strands/SWG	1500/42	1500/42
Resonance frequency (kHz)	85	85
Self-inductance (μH)	112.2	112.2
Compensating capacitor (nF)	31.22	31.22
Coil resistance (Ω)	1.5	1.5
Wire diameter (mm)	5.5	5.5
Inner radius (mm)	25	25
Outer radius (mm)	175	175
Shape	Circular	Circular

Table 11.5 DIBBC Simulation and Hardware Design Comparison

Parameter	Simulation Value	Hardware Value
Inductor	6 mH	5.5 mH
Filter Capacitor	1000 μF	2000 μF
Load	12 Ω	0–50 Ω
Battery	Lead Acid 120 V/165 Ah	Lead Acid 200 Ah (10Nos12 V/20 Ah)

Table 11.6 Specification of Solar PV in Solar Simulator

Parameter	Value
No. of Series/Parallel	2/3
Open-circuit voltage (V_{oc})	37.2 V
Short circuit current (I_{sc})	8.62 A
Maximum power condition	30.2 V/8.1 A
Maximum power (W)	213.5

Integrated dual input converter 291

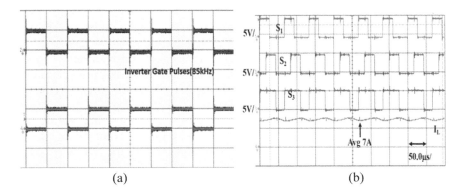

Figure 11.17 Gate pulse for (a) inverter switch (G1-G4), (b) DIBB converter switch (S1-S2).

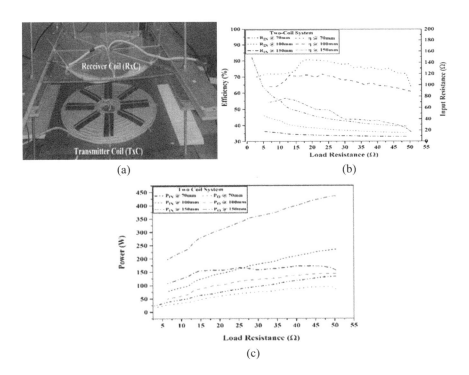

Figure 11.18 (a) Two-coil system arrangement. (b) Efficiency plots for the different transmission distance. (c) Power plots for different transmission distances.

receiver coil input voltage and current are shown in Figures 11.19 and 11.20. Also, input to the dual input converter from the wireless charger source after rectification and filtering is shown in Figure 11.21. Ripples in the transmitter voltage are observed due to incomplete compensation of the reactive power drawn by the coils. The nature of the transmitter current is observed to be peaky sinusoidal in one half due to the saturation from harmonic components as seen in transformers. Similarly, the receiver side voltage is not exactly square wave due to incomplete compensation coils. The load voltage shown in Figure 11.21 is measured across the filter capacitor connected across the WPT input terminals of the DIBBC. The rectifier output current passes a filter inductor and the flows to the filter capacitor and input to the DIBBC.

The two sources, WPT and solar PV or solar simulator, are maintained at 36 V and 60 V. The system is tested by inducing disturbance by varying source voltages of the DIBBC. The DIBBC is tested for four test conditions: V_{WPT} varying, V_{PV} varying, R_L varying and solar PV off. Inductor voltage (V_L) and inductor current (I_L) for the converter are shown in Figure 11.22. The test regions in Figure 11.23 show that the variations in solar PV and load side vary the output from the WPT source. When the load is increased,

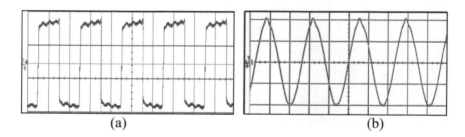

Figure 11.19 WPT system transmitter: (a) voltage (151 V_{rms}) and (b) current (I_{rms} = 6.9 A).

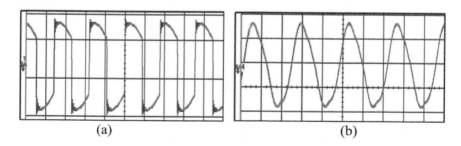

Figure 11.20 WPT system receiver: (a) voltage (66 V_{rms}) and (b) current (I_{rms} = 4.89 A).

Integrated dual input converter 293

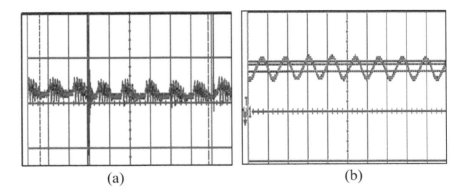

Figure 11.21 WPT system input to DIBBC: (a) voltage (62 V_{rms}) and (b) current (I_{rms} = 4.6 A).

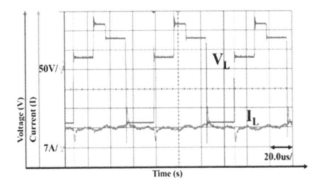

Figure 11.22 Integrated charging system inductor voltage (V_L) and inductor current (I_L).

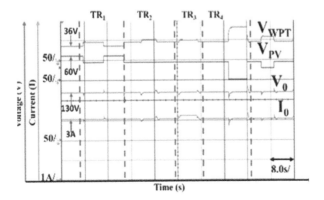

Figure 11.23 Charging system response under variable sources.

Table 11.7 Test Regions for Experimental Analysis of DIBBC

Test Region	Load	V_{PV}	V_{WPT}	V_o	I_o
TR_1	Constant	Decreased	Increases	Constant	Constant
TR_2	Constant	Constant	Decreased	Constant	Constant
TR_3	Increased	Constant	Increases	Constant	Increases
TR_4	Constant	OFF	Increases	Constant	Constant

i.e., test region 3, the solar PV output voltage is constant, whereas the WPT voltage increases. Details are tabulated in Table 11.5. Also, the WPT source is successful in maintaining the demand by the load in the absence of solar PV. In Figure 11.24, the DIBBC is operated as a single-source DC–DC converter, and output is constant in both cases.

The current drawn by DIBBC when operated with the WPT source is 12.5 A at 36 V and 3 A at 109.4 V. The efficiency of the dual-input converter is 87.3% when only WPT input is active and 72.2% when the only solar simulator is active. Thus, the efficiency of the DIBBC is high for high source voltages, as seen with the WPT source. To improve efficiency with solar PV, a high-voltage solar PV module should be used. During dual-source operation for output voltage of 130 V when input sources are set to 36 V and 60 V, the currents supplied by both sources are 3.9 A from the WPT source and 4.1 A from the solar simulator source. WPT source current (I_{WPT}), solar PV current (I_{PV}), load current and load voltage profile during the variation of load from 50 to 28 Ω are shown in Figure 11.25a. During the dual-source operation, the load is varying from 28 to 50 Ω. The efficiency plot is given in Figure 11.26a and the power plot in Figure 11.26b. Also, the battery charging current shown in Figure 11.25b is recorded for a 120 V/200 Ah battery stack. The charging current during constant voltage charging decreases from 4 to 1.5 A in the charging duration of 1000 seconds.

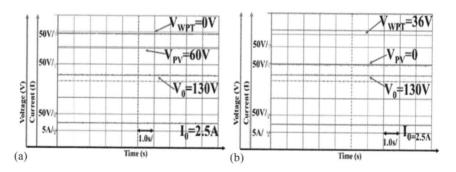

Figure 11.24 Single-source operation of DIBBC: (a) only solar PV active, (b) only WPT active.

Integrated dual input converter 295

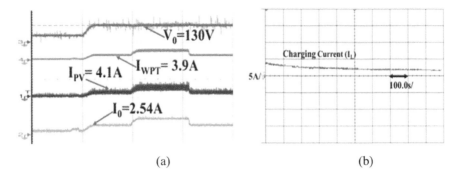

Figure 11.25 (a) DIBBC system load voltage and current profiles for varying load conditions. (b) Charging current under constant voltage charging mode.

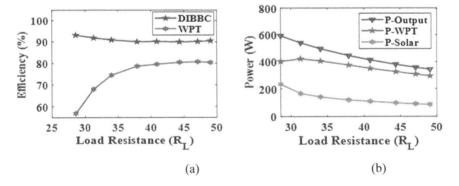

Figure 11.26 Charging system performance. (a) Efficiency of WPT and DIBBC. (b) Power output of DIBBC, input power from WPT and solar PV source.

The efficiency of the converter and WPT system measured for different load values is shown in Figure 11.26. The majority of the power to the load is supplied from the WPT system. Change in load resistance causes power flow from both the sources to change, while keeping the percentage of power supplied by the two sources unaltered. Also, it can be seen from the efficiency and power plots that the converter is more efficient when the difference between the power supplied from the two sources is lowest. This is observed at 28 Ω load and also at 50 Ω load. Thus, if both the sources supply equal power, the converter can perform more efficiently and can maintain high efficiency values for different loads as well. The average loss distribution pie chart for the dual input converter and WPT system is

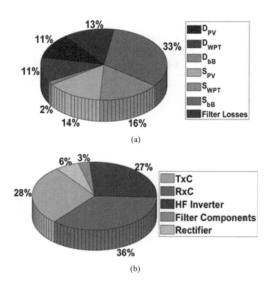

Figure 11.27 Loss distribution of (a) dual-input buck-boost converter and (b) wireless power transfer system.

shown in Figure 11.27. The losses in switches of the DIBB converter are switching losses, off-state leakage losses and conduction losses. S_{bB} contributes the majority of the switch losses, as it is on for a time period of $T = T_s(d_1 + d_2 + d_3)$. Thus, conduction losses are higher in the S_{bB}, which increases the overall losses. In diodes, the losses in D_{PV} and D_{WPT} are almost equal, and losses in D_{bB} are almost negligible. Filter inductor and capacitor losses contribute 13% of total losses. Thus, better design of filter components and PWM generation algorithm is required to further improve the efficiency of the converter. In the WPT system, the TxC, RxC and inverter are the major loss-contributing components. The highest loss occurs in the receiver coil, mainly due to harmonics and incomplete compensation. Thus, a detailed study to improve the coil fabrication and compensation topology is required to improve the efficiency of the WPT system. Also, the implementation of an effective PWM generation algorithm to reduce the harmonics will greatly contribute to further increasing the efficiency of the system.

11.7 Conclusion

This chapter study investigates the performance of a dual-input buck-boost converter (DIBBC) for renewable source integration with a wireless charger for EVs to reduce stress on the grid. On-board solar PV is the renewable energy source, and a WPT system is the wireless charger. A small-signal model of the DIBBC is derived, and further transfer functions for each switch S_1, S_2 and S_3 are derived. A PI controller for each switch is designed using the transfer function of the respective switch, and a control scheme is developed to operate the three switches to maintain the output voltage constant of the DIBBC during any disturbance. The simulation studies of the DIBBC are done in MATLAB Simulink. Simulation results show robust performance by the controller during disturbance injection into the two input sources. Hardware analysis of the DIBBC is done with a wireless charging system and solar simulator. The solar simulator is programmed to operate as a 60 V/25 A solar PV module. The efficiency of the DIBBC was observed to be always above 85% for the load range 28 to 50 Ω for an output voltage of 130 V. A detailed investigation of losses in the converter has also been discussed. Single-source operation of the dual-input converter gives low efficiency with the WPT system, as it supplies power at low voltage. Also, when the two sources are supplying almost equal power, the converter performance is further improved. The controller implemented for the developed DIBBC does not include the MPPT feature, and thus, the efficiency with an actual solar PV panel may be lower. Integrating the MPPT control technique with the proposed control scheme may improve the efficiency of the DIBBC as well as the solar PV performance. Further potential work on the proposed system includes the bidirectional operation of the converter for vehicle-to-grid using WPT and integration of MPPT in the control scheme.

References

1. Peerapat Vithayasrichareon, Graham Mills, and Iain F. MacGill. "Impact of electric vehicles and solar PV on future generation portfolio investment." *IEEE Transactions on Sustainable Energy* 6, no. 3 (2015): 899–908.
2. Samir M. Shariff, Mohammad Saad Alam, Furkan Ahmad, Yasser Rafat, M. Syed Jamil Asghar, and Saadullah Khan. "System design and realization of a solar-powered electric vehicle charging station." *IEEE Systems Journal* 14, no. 2 (2019): 2748–2758.
3. Mohammad Ekramul Kabir, Chadi Assi, Mosaddek Hossain Kamal Tushar, and Jun Yan. "Optimal scheduling of ev charging at a solar power-based charging station." *IEEE Systems Journal* 14, no. 3 (2020): 4221–4231.
4. Viet Thang Tran, Md Rabiul Islam, Kashem M. Muttaqi, and Danny Sutanto. "An efficient energy management approach for a solar-powered EV battery charging facility to support distribution grids." *IEEE Transactions on Industry Applications* 55, no. 6 (2019): 6517–6526.
5. Christian Schuss, Tapio Fabritius, Bernd Eichberger, and Timo Rahkonen. "Impacts on the output power of photovoltaics on top of electric and hybrid electric vehicles." *IEEE Transactions on Instrumentation and Measurement* 69, no. 5 (2019): 2449–2458.
6. Partha Sarathi Subudhi, Krithiga Subramanian, and Binu Ben Jose Dharmaian Retnam. "Wireless electric vehicle battery-charging system for solar-powered

298 *Power electronics for electric vehicles*

residential applications." *International Journal of Power and Energy Systems* 39, no. 3 (2019).

7. Kevin R. Mallon, Francis Assadian, and Bo Fu. "Analysis of on-board photovoltaics for a battery electric bus and their impact on battery lifespan." *Energies* 10, no. 7 (2017): 943.

8. Yihua Hu, Chun Gan, Wenping Cao, Youtong Fang, Stephen J. Finney, and Jianhua Wu. "Solar PV-powered SRM drive for EVs with flexible energy control functions." *IEEE Transactions on Industry Applications* 52, no. 4 (2016): 3357–3366.

9. Stefano De Pinto, Qian Lu, Pablo Camocardi, Christoforos Chatzikomis, Aldo Sorniotti, Domenico Ragonese, Gregorio Iuzzolino, Pietro Perlo, and Constantina Lekakou. "Electric vehicle driving range extension using photovoltaic panels." In *2016 IEEE Vehicle Power and Propulsion Conference (VPPC)*, pp. 1–6. IEEE, 2016.

10. W. A. N. G. Shuo, and D. G. Dorrell. "Loss analysis of circular wireless EV charging coupler." *IEEE Transactions on Magnetics MAG* 50, no. 11 (2014): 1–4.

11. Matjaz Rozman, Augustine Ikpehai, Bamidele Adebisi, Khaled M. Rabie, Haris Gacanin, Helen Ji, and Michael Fernando. "Smart wireless power transmission system for autonomous EV charging." *IEEE Access* 7 (2019): 112240–112248.

12. S. Chopra, and P. Bauer, "Driving range extension of EV with on-road contactless power transfer—A case study." *IEEE Transactions on Industrial Electronics* 60, no. 1 (2013): 329–338. https://doi.org/10.1109/TIE.2011 .2182011.

13. Partha Sarathi Subudhi, and S. Krithiga, "Wireless power transfer topologies used for static and dynamic charging of EV battery: A review." *International Journal of Emerging Electric Power Systems*, vol. 21, no. 1 (2020): 20190151. https://doi.org/10.1515/ijeeps-2019-0151.

14. C. Kalialakis, A. Collado, and A. Georgiadis, "Regulations and standards for wireless power transfer systems." In Nikoletseas S., Yang Y., and Georgiadis A. (eds.), *Wireless Power Transfer Algorithms, Technologies and Applications in Ad Hoc Communication Networks*. Springer, Cham, 2016. https://doi.org /10.1007/978-3-319-46810-5_7.

15. Sivaprasad Athikkal, Gangavarapu Guru Kumar, Kumaravel Sundaramoorthy, and Ashok Sankar. "Performance analysis of novel bridge type dual input DC–DC converters." *IEEE Access* 5 (2017): 15340–15353.

16. Sivaprasad Athikkal, Gangavarapu Guru Kumar, Kumaravel Sundaramoorthy, and Ashok Sankar. "A non-isolated bridge-type DC–DC converter for hybrid energy source integration." *IEEE Transactions on Industry Applications* 55, no. 4 (2019): 4033–4043.

17. Sivaprasad Athikkal, Kumaravel Sundaramoorthy, and Ashok Sankar. "Design, fabrication and performance analysis of a two input—Single output DC–DC converter." *Energies* 10, no. 9 (2017): 1410.

18. S. Kumaravel, G. Guru Kumar, Kuruva Veeranna, and V. Karthikeyan. "Novel non-isolated modified interleaved DC–DC converter to integrate ultracapacitor and battery sources for electric vehicle application." In *2018 20th National Power Systems Conference (NPSC)*, pp. 1–6. IEEE, 2018.

19. Ankit Kumar Singh, Manoj Badoni, and Yogesh N. Tatte. "A multifunctional solar PV and grid based on-board converter for electric vehicles." *IEEE Transactions on Vehicular Technology* 69, no. 4 (2020): 3717–3727.
20. D. K. Behera, I. Anand, B. Malakonda Reddy, and S. Senthilkumar, "A novel control scheme for a standalone solar PV system employing a multiport DC–DC converter." *2018 9th International Conference on Computing, Communication and Networking Technologies (ICCCNT)*, Bengaluru, India, pp. 1–6, 2018, https://doi.org/10.1109/ICCCNT.2018.8494101.
21. P. S. R. Nayak, K. Kamalapathi, N. Laxman, and V. K. Tyagi, "Design and simulation of buck-boost type dual input DC–DC converter for battery charging application in electric vehicle." *2021 International Conference on Sustainable Energy and Future Electric Transportation (SEFET)*, Hyderabad, India, pp. 1–6, 2021. https://doi.org/10.1109/SeFet48154.2021.9375658.
22. Lalit Kumar, and Shailendra Jain. "Multiple-input DC/DC converter topology for hybrid energy system." *IET Power Electronics* 6, no. 8 (2013): 1483–1501.
23. X. H. Nguyen, and M. P. Nguyen. "Mathematical modeling of photovoltaic cell/module/arrays with tags in Matlab/Simulink." *Environmental Systems Research* 4 (2015): 24.

Index

Acceleration period 10
ADC 74
Advance timer 74
Ampere's law 177
Arbitrary angular speed 253
Asymmetric H-Bridge 203
Asymmetrical clamped mode control 175
Asymmetrical duty cycle control 179
Attenuated voltage 76
Average model 47
Axle balancing mode 12

Bandwidth 53
Battery charging system 103
Battery electric vehicles 2, 218
Battery EVs 27
Battery interface 37
Battery pack 93
Battery storage system 122, 217
Battery voltage 158, 184
Battery-operated electric vehicles 117
Bidirectional converter 4, 93
Bidirectional energy converter 7
Bidirectional OBCs 36
Bifurcation 185
Bode plot 181
Boost factor 226
Boost inductor 40
Braking mode 10
Bridgeless PFC 38
Bridge-type PFC 38
Brushless dc motor 5, 192
Buck-boost 102

Capacitive Power Transfer 139
CC-CV control 71
Charge-de-Move 30
Charging algorithm 72

Charging current 295
Charging protocol 31
Circulating current 68
Closed loop 186
Common-mode noise 41
Comparative analysis 54
Compensating capacitor 290
Complex hybrid drive system 9
Conduction path 40
Conductivity of Coil Material 161
Constant current (CC) charging mode 51
Constant current 138
Constant current mode 186
Constant voltage 51, 138
Constant voltage mode 186
Continuous conduction 38
Control algorithm 109
Control pilot line 36
Controller design 47
Controller's gain 53
Converter prototype 70
Coupling Coefficient 145
C-rate 184
C-rating 73
Critical K 158
Crossover frequency 52
Current Source Inverter 151

DAB 45
DC–DC converter 93
Dead time 77
Dedicated OBCs 29
Delta-Q OBC 45
Describing function 183
Digital communication 31
DIHB-SAFBR DC–DC converter 97
Direct average torque control 200
Direct Current 145

302 *Index*

Direct instantaneous torque control 200
Direct torque and flux control 200
Discontinuous conduction 38
Disturbance injection 296
Drivetrains 6
Dual active bridge currents 105
Dual axle propulsion system 9
Dual-input buck-boost converter 296
Dual-input converters 273
Dynamic Wireless Charging 140

Efficiency 157, 294
Electric Field 145
Electric mobility 119
Electric propulsion 5
Electric traction motors 5
Electric vehicle battery charging
 system 113
Electric vehicles (EVs) 93, 137, 175
Electric-ICE hybrid 14
Electromagnetic compatibility 30
Electromagnetic interference 146, 179
Electromagnetic torque 255
Electromotive force 197, 246
Energy management system (EMS) 119
Energy storage 4
Energy storage systems (ESS) 93,
 118, 217
Energy storage unit 6
Environmental pollution 5
Error voltage 52
EV supply equipment 27
EVSE 36
Extended describing function 180

Faraday's law 177
Field-oriented control 246
Finite element analysis 138
Floating 41
Foreign object detection 168
Freewheeling 66
Frequency modulation 61
Frequency modulator 80
Front-end 40
Fuel cell 3
Fuel cell electric vehicle 2, 218
Fuel cell–based hybrid 14
Fuel efficiency 9
Full hybrid vehicle 13
Fundamental harmonic analysis 62

G2V 43
Gain margin (GM) 285
Gallium nitride 45, 168

Galvanic isolation 37, 113
Gate pulse 51
Gauss's law 177
Green generation systems 93
Grid power supplies 93
Grid-to-vehicle 154

Half-bridge bidirectional DC–DC 113
Hall current sensors 77
Hall voltage sensor 74
Harmonic approximation 182
Harmonic balance 183
Harmonic free current 51
H-Bridge 153
High current spikes 65
High frequency AC 145
High gain sepic-based ZSI 219
Higher-order harmonics 61
High-frequency alternating current 273
High-frequency transformer 42, 43
High-speed driving mode 11
Holding time 46
Hybrid electric vehicle 1, 27
Hybrid power transfer 139
Hybrid renewable energy schemes 95
Hydrogen gas 3

Inductance ratio 64
Induction motors 192
Inductive coils 176
Inductive power transfer 138
Inductive wireless power transfer 177
Inductor-Capacitor-Inductor 149
Inner current loop 51
Input current 37
Institute of Electrical and Electronics
 Engineers 143
Insulated gate bipolar transistor 227
Integrated leakage inductance 88
Integrated magnetics 61
Integrated OBCs 29
Interleaved boost PFC 28
Internal combustion 192
Internal combustion engine 1, 218
International Commission on Non-
 Ionizing Radiation Protection 143
Irradiance 275
Isolated and non-isolated topologies 94
Isolated OBCs 37
Iteration 68

K-factor method 180
Kirchhoff current law 223
Kirchhoff voltage law 155, 178, 223

Index 303

Korea Advanced Institute of Science and Technology 141

Leakage inductance 61
Light load mode 11
Line regulation 86
Load torque 254
Loss distribution 296
Lower saturation limits 73
Low-frequency AC 150
Low-power 39

Magnetizing inductance 66, 254
MATLAB/Simulink 51
Matrix converter 43, 151
Maximum constant boost PWM 235
Maximum power condition 290
Maximum power point tracking 248
Medium hybrid electric vehicles 13
Microcontroller 74
Mild hybrid vehicles 12
Minimum load condition 10
Model reference adaptive control 246
Modulation index 236
MPPT (Maximum Power Point Tracking) 93
Multi-input converter (MIC)/ multiport DC–DC topologies 94
Mutual inductance 276

Nano grid 124
Nominal voltage 73
Non-isolated OBCs 37
Normal driving mode 10
Normalized switching frequency 64

Oak Ridge National Laboratory 141
Off-board chargers 29
OnLine EV 143
On-board chargers 28, 118
On-board solar PV-aided EV charging 272
Outer voltage loop 51
Output filter 51

Parallel hybrid vehicle 7
Parallel-parallel 149
Parallel resonant 42
Parallel-series 149
Parasitic capacitance 102
Peak current 48
Permanent magnet brushless DC motor 246
Permanent magnet motors 192

Permanent magnet synchronous motor 5, 192, 246
Permeability of coil material 161
PFM control 71
Phase locked loop 246
Phase margin (PM) 181, 285
Phase shift control 179
Planetary gear 8
Plug-in HEVs 2, 27
Pole pair 255
Positive sequence flux 259
Power control unit 6
Power delivery 65
Power density 45
Power electronic converters 93, 138
Power factor correction 145
Power plots 291
Power transfer to the load 157
Powertrain 13
Primary and Secondary Coil Quality Factors 157
Primary side 45
Primary voltage 49
Proportion integral 248
Proportional controller 164
Proportional-integral (PI) 104
Proportional-integral (PI) controller 52
Propulsion torque 27
Proximity pilot 30
PSFB converter 45
Public EV charging station 125
Pulsating DC 39
Pulse width modulation 164
Pulse width modulation control 71
PV, wind, and biomass 95
PV-based charging stations 271

Quality factor 64, 185
Quasi ZSI 219

Receiver coil 290
Receiver 178
Rechargeable battery 3
Rectification 292
Reference current 52
Reflected load resistance 64
Regenerative braking 4, 27, 118
Renewable energy sources (RES) 93
Renewable-powered charging station 271
Resonance frequency 277
Resonant 42
Resonant capacitor 70
Resonant frequency 64, 180

304 *Index*

Reverse recovery losses 65
Ripple current 41, 225
Ripple-free voltage 51
Rooftop solar photovoltaic (PV) systems 271
Rotor fluxes 254
Rotor position error 257

Sampling time 76
Saturation limit 71
Sawtooth signal 51
Secondary current 79
Secondary side variables 62
Secondary voltage 49
Second-gen volt charger 46
Self-inductance 278
Semi-bridgeless PFC 41
Semi-onboard/ off-board chargers 29
Semi-ZSI 226
Series hybrid configuration 6
Series resonance inductance 61
Series resonant network 275
Series-parallel hybrid vehicle system 8
Series-parallel resonant converter 43
Series-parallel 149
Series-series 149
Series-series compensation 272
Series-series resonant converter 179
Short circuit current 275
Silicon carbide 168
Simulation waveforms 82
Sinusoidal current 54
Sinusoidal ripple current 151
Skin depth 161
Sliding mode observer 246
Small signal model 181
Small-scale DC charging system 121
Small-signal model 283
Society of Automotive Engineers 30, 143
Soft commutation 79
Soft switching 176
Soft-switching circuits 42
Solar photovoltaic 93, 245
Solar-powered charging 119
Solar-powered electric vehicles 245
SPARTAN-6 FPGA board 290
Spiral circular coil 176
Square wave inverter 42
Starting mode 11

State equations 47
State of charge 119
State-space model 47
Stator flux linkages 254
Step-up operation 81
Stopping mode 10
Switched reluctance motor 5
Synchronous reference frame 251

Three port converter 96
Timer register 78
Torque sharing functions 200
Torque–speed characteristics 7
Total harmonic distortion 28, 145
Traction motor 29
Transfer function 62
Transformation ratio 49
Transient operation 84
Transmitter 178
Transmitter coil 290
Trans-motor 8

Ultra-capacitor 3
Uncontrolled rectifier 38
Unidirectional energy converter 7
Unidirectional OBCs 36
Universal input voltage 46

Vehicle-to-grid (V2G) mode 117
Vehicle-to-grid 43, 154
Voltage divider 42
Voltage gain 69, 157
Voltage lift cell 228
Voltage regulation 67
Voltage source inverter 150, 247
Volt-Ampere 145

Wideband gap devices 57
Wireless Advanced Vehicle Electrification 141
Wireless battery charging 175
Wireless chargers 29
Wireless charging stations 272
Wireless power transfer 137, 272
Wound inductor 75

Zero current switching 45, 175
Zero phase angle 156
Zero voltage switching 61, 96, 175
Z-source inverter 217